Synthesis Lectures on Mathematics & Statistics

Series Editor

Steven G. Krantz, Department of Mathematics, Washington University, Saint Louis, MO, USA

This series includes titles in applied mathematics and statistics for cross-disciplinary STEM professionals, educators, researchers, and students. The series focuses on new and traditional techniques to develop mathematical knowledge and skills, an understanding of core mathematical reasoning, and the ability to utilize data in specific applications.

Robert Hirsch

Statistical Hypothesis Testing with Microsoft® Office Excel®

 Springer

Robert Hirsch
Overland Park, KS, USA

ISSN 1938-1743   ISSN 1938-1751   (electronic)
Synthesis Lectures on Mathematics & Statistics
ISBN 978-3-031-04204-1   ISBN 978-3-031-04202-7   (eBook)
https://doi.org/10.1007/978-3-031-04202-7

This Springer imprint is published by the registered company Springer Nature Switzerland AG
The registered company address is: Gewerbestrasse 11, 6330 Cham, Switzerland

Preface

If you need to interpret the results of statistical tests, want to be able to analyze your own data, or plan a study without having to consult with a statistician, this text is for you. It is written for the researcher who does not want to take a statistics course but needs to understand what the statistician does. It is largely nonmathematical except where absolutely necessary. It does not assume the reader knows much about statistics except what a mean and proportion are. To get the most out of the text, you should have access to Microsoft® Office Excel® but you need not be an expert on its use.

In Chap. 1 you will learn what a P-value is, where it comes from, and how to interpret its value. You will also learn about the type of errors that can be made when drawing conclusions based on the P-value. You will become familiar with two approaches: testing hypotheses and their advantages and disadvantages. You will learn how to use these approaches to interpret a collection of several statistical tests in the same study.

Chapter 2 explains how samples can be taken from the population in a way that creates an unbiased and analyzable dataset. The steps in the sampling process are examined. Then, the methods of drawing the sample are explained with their advantages and disadvantages.

In Chap. 3 you will learn how to choose a statistical test that is appropriate for a given set of data. Then, you will learn about the most common statistical tests, how to use Excel to perform them, and how to interpret the results by using them.

Chapter 4 discusses what you can do if you don't want to wait until a study gets to its planned end before you start looking at the data and draw conclusions about what will happen if the study is allowed to go to completion. It discusses two approaches, both of which can be used to decide to stop a study early if the results look promising. One can also stop a study early if the results do not look promising.

Chapter 5 tells you how to determine the size of a planned sample that allows it to be big enough to see important relationships without being bigger than required. Methods are given for nominal and continuous variables and for datasets with and without independent variables.

Overland Park, KS, USA   Robert Hirsch

Acknowledgments I appreciate my clients who have shared their frustration in trying to understand statistical hypothesis testing and their desire to be able to perform some statistical procedures without having to involve a statistician.

Notices

The examples in this book are based on fictitious data and should not be taken as a reflection of real relationships. Those data have been created only to illustrate statistical principles.

Contents

Logic of Hypothesis Testing

1

Abstract

This chapter begins by describing the hypothesis statistics is designed to test. That hypothesis, known as the null hypothesis, states that things do not differ or there is no association between measurements. If that hypothesis is rejected, we conclude that there are differences or associations. Decisions to reject the null hypothesis are based on P-values. The chapter describes the origin and interpretation of P-values. It also discusses errors that can occur in interpretation of the P-value and how to control them. This discussion addresses the classical or frequentist approach to hypothesis testing. The Bayesian approach takes things further, allowing determination of the probability that the null hypothesis is false given frequentist methods have resulted in rejection of the null hypothesis. Both approaches are applied to the situation in which a study includes several hypothesis tests.

1.1 Classical (Frequentist) Approach

The *classical* or *frequentist* approach to hypothesis is the one used most often and is taught in most introductory statistics texts. We will begin by understanding the logic behind this approach to statistical hypothesis testing.

1.1.1 Statistical Hypotheses and Conclusions

Suppose researchers are comparing a new antiviral treatment to the standard treatment as to how well they keep infected persons from hospitalization. Those researchers hypothesize that the new treatment is better than the standard treatment and they want a statistical

© The Author(s), under exclusive license to Springer Nature Switzerland AG 2022   1
R. Hirsch, *Statistical Hypothesis Testing with Microsoft® Office Excel®*,
Synthesis Lectures on Mathematics & Statistics.
https://doi.org/10.1007/978-3-031-04202-7_1

analysis of their data to support that hypothesis. The way statistics can do that is by eliminating chance as a likely reason for an observed difference between the two treatments.[1] To do that, statistics calculates the probability of observing the observed difference if a hypothesis about the difference is true. Unfortunately, this probability cannot be calculated using the researchers' hypothesis. To calculate the probability, the hypothesis must make a specific statement about the difference between the treatments. The researchers' hypothesis just states that the new treatment is better but does not specify how much better. To make a specific statement about the difference between the treatments, statistics uses a hypothesis that the researchers believe is not true. That hypothesis is that there is no difference between the treatments. That is a specific statement that the difference is equal to zero in the population. This statistical hypothesis is still useful to the researchers, because if it can be eliminated as a likely explanation for the observed difference, they can, through the process of elimination, conclude that there is a difference not due to chance.

This hypothesis used by statistics is called the *null hypothesis*. The null hypothesis is a statement that there is no difference or there is no relationship between measurements in the population from which the sample was taken. Statistics tests the null hypothesis by calculating the probability of getting the observed or a more extreme difference if the null hypothesis is true. This probability is called the *P-value*. If the *P*-value is small enough, the null hypothesis is rejected. The most used value for the *P*-value to reject the null hypothesis is 0.05 (or 5%). If the chance of getting the observed difference or more extreme assuming the null hypothesis is true is 5% or less, we reject the null hypothesis. If the chance of getting the observed or more extreme difference is greater than 5%, we fail to reject the null hypothesis.[2]

If we reject the null hypothesis as an explanation for the observed difference, what do we conclude? This is specified by the *alternative hypothesis*. The alternative hypothesis is not tested. Rather, it is embraced as a reflection of truth only through the process of eliminating the null hypothesis. Since it is accepted as truth through the process of elimination of the null hypothesis, the alternative hypothesis must include all possibilities except that stated in the null hypothesis. For the example of the antiviral treatment, the most likely alternative hypothesis is that there is a difference between the treatments in the population. This is called a *two-tailed* (or *two-sided*) alternative hypothesis. A two-tailed alternative hypothesis would include both the new treatment being better and the new treatment being worse than the standard treatment. If it is impossible for the new

---

[1] Another possible explanation for an observed difference is that there is a bias in the design of the study. For instance, one group might be sicker than the other group. If both chance and bias can be eliminated as likely explanations, the observed difference must be reflecting a causal relationship.

[2] Note that we do not accept the null hypothesis as true if the *P*-value is greater than 0.05. There is no rule of thumb for how large the *P*-value needs to be to believe the null hypothesis is true in the frequentist approach.

treatment to be worse than the standard treatment, we can use a *one-tailed* (or *one-sided*) alternative hypothesis.[3] Then elimination of the null hypothesis would allow us to conclude that the new treatment is better than the standard treatment in the population.

1.1.2   Errors in Hypothesis Testing

Since we are relying on probabilities (*P*-values) in drawing conclusions, something can go wrong. For example, we could reject the null hypothesis even though the null hypothesis is true. This is called a *type I error*. The chance of making a type I error is determined by our choice for how small the *P*-value must be to reject the null hypothesis. As stated previously, the most common value is 0.05 (5%). This value is called *alpha*. An alpha of 0.05 means that we reject the null hypothesis when the *P*-value is equal to or less than 0.05. It also means we have a 5% chance of making a type I error.

Another possible error would be to accept the null hypothesis as true when it is not true. This is called a *type II error*. The chance of making a type II error is called *beta*. Beta is the probability of accepting the null hypothesis as true given that the alternative hypothesis is true. Beta cannot be calculated because the alternative hypothesis is not a specific statement about the difference or relationship. Since we don't know the chance of making a type II error, we avoid risking making it. We do that by not concluding that the null hypothesis is true when we cannot reject it. Instead, we "fail to reject" the null hypothesis. This is tantamount to not drawing a conclusion from the results of hypothesis testing when we cannot reject the null hypothesis.

The complement of beta (1-beta) is called the *statistical power*. Statistical power is the probability of rejecting a false null hypothesis. It is calculated when planning a study. To calculate it, statisticians need to specify a specific value for the alternative hypothesis. Then, the calculated statistical power is relevant only when the specified value of the difference (or relationship) is true. Statisticians usually use a difference that is the threshold of interest. For example, suppose the researchers think that the new treatment would be worthwhile only if it is as least 2% better than the standard treatment. Then, statisticians would use a difference of 2% to calculate the statistical power when planning the study. This is discussed further in Chap. 5.

1.1.3   Calculating *P*-Values

To calculate a *P*-value we need to think about the distribution of all possible values for whatever we are estimating. For example, if we are interested in differences in the chance

---

[3] Note that to use a one-tailed alternative hypothesis, deviations from the null hypothesis must be possible in only one direction. It is not enough to say we do not think that deviations in both directions will occur or to say we are only interested in deviations in one direction.

Fig. 1.1 The sampling
distribution for the difference
in the chance of hospitalization
under the null hypothesis that
that difference is equal to zero
in the population

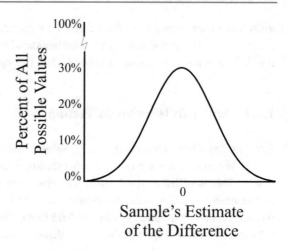

of hospitalization between the new treatment and the standard treatment, we need to think
about the distribution of all possible differences. This distribution tells us how frequently
each value of the difference would occur if we were to run the experiment many, many
times on different samples of patients. This is called the *sampling distribution*.[4]

To calculate a *P*-value, we must assume that the null hypothesis is true. This is reflected
in the sampling distribution by the null value being the most frequent value in the distribution. This sampling distribution is called the *null distribution*. Figure 1.1 shows the
null distribution for the study of a new antiviral treatment.

On that null distribution, the observed difference, and its negative value (for a
two-tailed alternative hypothesis) are considered. This defines the "tails" of the null
distribution. Suppose the researchers observe a difference of 5% in the chance of
hospitalization. This null distribution is illustrated in Fig. 1.2.

The *P*-value is the proportion of the null distribution in the tails. This is illustrated in
Fig. 1.3.

1.1.4 Interpreting *P*-Values

To interpret a *P*-value it is compared to alpha, the selected probability of making a type
I error. As previously stated, the most common value of alpha is 0.05. If the *P*-value
is less than or equal to alpha, we reject the null hypothesis and, through the process of
elimination, accept the alternative hypothesis. In this case, we say the result is *statistically
significant*.

---

[4] I have drawn the sampling distribution as a bell-shaped curve. Sampling distributions tend to be
bell-shaped regardless of the shape of the distribution of data, if the sample is large enough. This
important principle is called the *central limit theorem*.

Fig. 1.2 The tails of the null distribution for a two-tailed alternative hypothesis if the researchers observe a difference of 5% in the chance of hospitalization

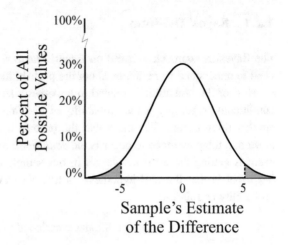

Fig. 1.3 The *P*-value as the proportion of the null distribution in the tails of that distribution

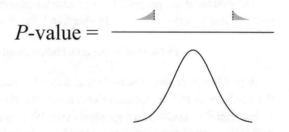

What about the numeric magnitude of the *P*-value? What does a *P*-value of 0.01 tell us? Other than rejecting the null hypothesis, it just tells us that the sample we got would happen 1% of the time if the null hypothesis were true. It does not tell us the chance that the null hypothesis is true. Instead, the *P*-value assumes that the null hypothesis is true. It has no relevance if the null hypothesis is false.

1.2   Bayesian Approach

The approach described in the previous chapter is the classical approach to hypothesis testing. It is often called the frequentist approach. Another approach is the *Bayesian approach*. An advantage of the Bayesian approach is that it can allow us to calculate the probability that the null hypothesis is true given our observations.

1.2.1 Bayes' Theorem

The Bayesian approach is based on *Bayes' Theorem*. To understand Bayes' theorem, we need to understand some things about the probabilities related to hypothesis testing.

Most of the probabilities related to hypothesis testing are *conditional probabilities*. A conditional probability is the probability of one thing occurring under the condition that another thing occurs. The thing that the probability addresses is called the *conditional event* and thing assumed to occur is the *conditioning event*. For a *P*-value, the conditional event is getting the estimate we got in our sample or more extreme and the conditioning event is that the null hypothesis is true. In shorthand, we can write a conditional probability as follows:

$$p(\text{conditional event}|\text{conditioning event}) \tag{1.1}$$

The *P*-value is the probability of getting the observed difference or more extreme given that the null hypothesis is true. In shorthand, the *P*-value is:

$$P\text{-value} = p(\text{observed results}|\text{null hypothesis is true}) \tag{1.2}$$

A problem with the *P*-value is that it is a backwards probability. By that I mean that the conditional and conditioning events are the reverse of what we would like to have to interpret the results of a hypothesis test. What we would like to have is a conditional probability called the *posterior probability*.[5] The posterior probability tells us the probability that the null hypothesis is true given the observed estimate or more extreme. In shorthand, we can write the posterior probability as follows:

$$\text{Posterior} = p(\text{null hypothesis is true}|\text{observed results}) \tag{1.3}$$

In the Bayesian approach to hypothesis testing, there is a third probability we need to consider. That is the *prior probability*. The prior probability is the probability that the null hypothesis is true before we know that results of the statistical analysis.[6] It is not a conditional probability. In shorthand, the prior probability is:

$$\text{Prior} = p(\text{null hypothesis is true}) \tag{1.4}$$

Finally, in the Bayesian approach, we need to consider the statistical power. Statistical power is the probability of getting the observed estimate or more extreme given that the null hypothesis is false.[7] In shorthand, statistical power is:

---

[5] We call this the posterior probability because it is the probability of the null hypothesis being true after we know the results of the statistical analysis.

[6] This is often a difficult probability to which to assign a value. There is no analysis that helps. Rather, the research must assign a value based on a subjective guess.

[7] We will learn more about statistical power in Chap. 5.

$$\text{Power} = (\text{observed results}|\text{null hypothesis is false}) \tag{1.5}$$

The way in which we can interchange the conditional and conditioning events in a conditional probability is by using Bayes' theorem. In terms of the P-value, we can use Bayes' theorem to get the probability that the null hypothesis is true given the observed difference or more extreme. That calculation is as follows:

$$\text{Posterior} = \frac{P\text{-value} \cdot \text{Prior}}{[P\text{-value} \cdot \text{Prior}] + [\text{Power} \cdot (1 - \text{Prior})]} \tag{1.6}$$

Now, let us look at example of using Bayes' Theorem to interpret the results of a statistical hypothesis test.

Example 1.1 The researchers who compared a new antiviral medication to the standard treatment sent their data to a statistician for analysis. In the statistician's report, she says the difference in chance of hospitalization is 5% and, testing the null hypothesis that the difference is equal to zero in the population, the P-value is 0.02. Further, the statistician says that the study has a power of 0.80 to detect a difference of 5%. Interpret these results.

Using the frequentist approach, the researchers observe that the P-value is less than 0.05, thus they reject the null hypothesis and, through the process of elimination, accept the alternative hypothesis that the difference is not equal to zero. This is the limit of what they can do using the frequentist approach.

To get more information about the null hypothesis, the researchers decide to use the Bayesian approach. In preparation, they consider what they thought was the chance that the null hypothesis was true before the study was done. Since animal studies were very promising, they believed there was a very good chance of rejecting the null hypothesis before they did their study on patients. Thus, they believe there was a low chance that the null hypothesis is true. They pick a probability of 0.1 to represent that low probability.

Now, the researchers use Eq. 1.6 to calculate the posterior probability that the null hypothesis is true given that the null hypothesis has been rejected.

$$\text{Posterior} = \frac{0.02 \cdot 0.1}{[0.02 \cdot 0.1] + [0.80 \cdot 0.9]} = 0.00277$$

So, the probability that the null hypothesis is true given the observe difference is about 0.003. Or that can be expressed as the probability that the null hypothesis is false (1–0.00277 = 0.99723). That implies the researchers can be more than 99% confident that the difference in chance of hospitalization is not the same for the two treatments given the observed difference of 5%.

Because determination of the prior probability is a guess, other values can be considered to see how this affects the conclusion. This is illustrated in the next example.

Example 1.2 In Example 1.1, the researchers choose a prior probability of 0.1 to represent the belief that the null hypothesis that there is no difference between the treatments was unlikely to be true. To evaluate the effect of that guess on the conclusions, they tried other values for the prior probability.

The following table summarizes the results of considering other values for the prior probability.

Prior	Posterior
0.05	0.001314
0.10	0.002770
0.20	0.006211
0.30	0.010601
0.40	0.016393

All those posterior probabilities are small, so it does not matter what value they choose to represent their belief that the null hypothesis is unlikely to be true. The conclusion remains the same.

1.2.2 Multiple Hypotheses

Sometimes a study includes several hypothesis tests. One type of study in which this occurs is a study that is searching for risk factors for a disease. Generally, several risk factors are investigated. Each risk factor receives its own hypothesis test testing the null hypothesis that there is no difference in the risk of disease for people with the risk factor compared to people without the risk factor. In this circumstance a study can contain several hypothesis tests each with its own *P*-value.

A problem with this type of study is that there can be a high probability that at least one hypothesis test is statistically significant even though its corresponding risk factor is not really a characteristic that is associated with the disease. In other words, there can be a high probability of at least one type I error. This is known as the *multiple comparison problem*.

The chance of at least one type I error in a collection of hypothesis tests make up its *experiment-wise* type I error rate. In contrast, the chance of a type I error from a single hypothesis test is the *test-wise* type I error rate. The test-wise type I error rate is determined by our choice of alpha, the value we use to determine whether we reject the null hypothesis. The usual value of the test-wise alpha is 0.05.

The experiment-wise alpha can be substantially greater than the test-wise alpha. For instance, if we perform five hypothesis tests with a test-wise alpha of 0.05, the experiment-wise alpha is 0.2262. Most studies of this type have more than five risk factors considered. Twenty is not unusual. With twenty risk factors, the experiment-wise alpha is 0.6415. Many believe that this is a concern, and something should be done to control the experiment-wise type I error rate.

The frequentist approach to this problem is simple. That solution is to use a smaller value of test-wise alpha so that the experiment-wise alpha is 0.05. A way to do that is to divide 0.05 by the number of tests. This is known as the *Bonferroni correction*. So, for ten tests, the value of alpha to use to interpret each test is 0.005. Then, the experiment-wise alpha is 0.04889, very close to 0.05. It works!

There is a serious drawback to this method. That is adjusting the test-wise alpha to a smaller value increases the chance of failing to reject a false null hypothesis. In other words, this approach decreases the statistical power. This can cause us to miss detecting real relationships that we would have seen if we had not used a smaller test-wise alpha.

The Bayesian approach to this problem does not reduce statistical power. In this approach, we assign prior probabilities to each of the hypothesis tests. In the example of a study examining risk factors for a disease, this means we distinguish among the risk factors, identifying those that have a high biologic likelihood of being true predictors of the disease distinguishing them from the risk factors that have a low biologic likelihood of being true predictors of the disease. Then smaller *P*-values will be required to obtain an acceptable posterior probability for those low likelihood risk factors than the *P*-values required to obtain an acceptable posterior probability for the high likelihood risk factors.

Example 1.3 Suppose we are interested in risk factors for pancreatic cancer. We perform a study in which we evaluate 30 characteristics and obtain 5 statistically significant (i.e., with *P*-values less than or equal to 0.05) results. These are listed in the following table.

Risk factor	*P*-value
Family history	0.015
Smoking	0.001
High alcohol consumption	0.019
High coffee consumption	0.027
Obesity	0.011

Let us interpret these results.

First let us use the frequentist approach. To interpret these *P*-values, we compare them to the adjusted value of alpha. We get that adjusted value of alpha by dividing 0.05 by the number of tests (30). The adjusted alpha is 0.00167. When we use that adjusted alpha, only one characteristic (smoking) is statistically significant. We fail to reject the null hypothesis for the other characteristics.

Now, let us use the Bayesian approach. Before we did the study, we thought family history, smoking, high alcohol consumption, and obesity were likely to be real risk factors. Let us say a value of 80% represents the likelihood that we would be able to reject the null hypothesis for these characteristics. High coffee consumption, however, was not expected to be a real risk factor. Let us say that 30% represents the likelihood that we would be able to reject the null hypothesis for high coffee consumption. When planning the study, the sample size was selected to give us 90% power to detect a clinically relevant difference.

With that information, we can use Eq. 1.6 to calculate posterior probabilities that the null hypothesis of no difference in risk is true. The following table summarizes those calculations.

Risk factor	P-value	Prior	Posterior
Family history	0.015	0.2	0.0042
Smoking	0.001	0.2	0.0003
High alcohol consumption	0.019	0.2	0.0053
High coffee consumption	0.027	0.7	0.0654
Obesity	0.011	0.2	0.0030

Using a value of 0.05 or less for the posterior probability to consider the result to be statistically significant, all except high coffee consumption have a small enough posterior probability for us to believe that the null hypothesis is not true. In other words, we conclude that they are true risk factors. The evidence is just not strong enough to overcome the high prior probability that the null hypothesis is true for high coffee consumption.

Sampling

2

Abstract

Hypothesis testing involves examination of a sample to say something about the population from which the sample was drawn. An assumption of hypothesis testing is that the sample is representative of the population, at least on the average. This chapter describes the sampling process going from elements in the population to observational units in the sample. There are several methods for selecting the sample. The best of these is probability sampling that leads to different kinds of random samples. Sampling is usually a multistage process and can involve different sampling methods at each stage.

In statistical hypothesis testing, we examine the sample to draw conclusions about the population. Our ability to do that depends on how the sample was taken. Regardless of how sophisticated the method of analyzing data in the sample an improperly taken sample will preclude accurate conclusions about the population. Thus, it is important that we understand the basic issues in sampling. The universal assumption of all statistical procedures is that the sample be representative of the population for which we wish to test hypotheses. In this chapter, we will learn that there are several ways in which we can satisfy this requirement.

© The Author(s), under exclusive license to Springer Nature Switzerland AG 2022 11
R. Hirsch, *Statistical Hypothesis Testing with Microsoft* ® *Office Excel* ®,
Synthesis Lectures on Mathematics & Statistics.
https://doi.org/10.1007/978-3-031-04202-7_2

2.1   Taking Samples

The need for statistics in research stems from the fact that we examine samples so that we can draw inferences about populations from which the samples were drawn. From a statistical point of view, our interest in applying inferences is limited to the population from which the sample was drawn. That population is called the *sampled population*.[1] Each time we attempt to apply the results of analysis of a sample to the sampled population, we assume that the sample has been drawn in such a way that it is representative of the values in that population, at least on the average. This assumption, therefore, is universally part of every statistical procedure.

Before we discuss the various ways in which a sample can be taken so that it is representative of the population, however, we need to become familiar with the terminology that is used in sampling theory. The smallest units that are of interest in sampling from a population are the *elements* in the population. The elements are the units for which information is sought for estimation or inference. In research, persons are the most common elements, but other elements might be animals, assays, cultures, etc. The way to identify the elements is to ask yourself, "For what units in the population do I want to draw inferences?".

The units that might be selected to be in a sample are called the *sampling units*. Analysis is easiest when each sampling unit contains one element, but this is not always the case. For example, suppose we were interested in estimating serum cholesterol levels among adults in a certain community. The elements are adults in this community. If we were to draw our sample by selecting individuals from this population, the sampling units would also be adults in the community. It might be easier, however, for us to draw a sample of households in the community and to measure the serum cholesterol levels for all the adults in a selected household. If we were to sample the population in this way many of the sampling units (households) would contain more than one element (adults).

When sampling units contain one element, we are doing *elemental sampling*. When sampling units contain more than one element, we are doing *cluster sampling*. Most statistical procedures assume that we have an elemental sample.

Sometimes a sampling unit consists of multiple measurements of the data of interest for each element at a particular point in time. For example, in measuring serum cholesterol levels we might make determinations in triplicate and then use the mean of the three determinations as the serum cholesterol level for an individual. The reasoning behind this strategy is that each determination is made with a certain degree of imprecision and the imprecision is reduced by repeating the determination and taking the mean of the

---

[1] We are often interested in applying inferences to populations that are like, but not the same as, the sampled population. This process of applying estimates or inferences to populations other than the one from which the sample was drawn is called *extrapolation*. The populations to which we extrapolate inferences is called the *target population*. The appropriateness of extrapolation depends on the similarity of the sampled and target populations.

determinations.[2] In fact, what we have done is to obtain a sample from a population in which the sampling units contain more than one element.

When the sampling units contain more than one element, either of two approaches might be taken. One approach is to use a method of analysis that takes this into account. The other approach is to allow the sampling units to redefine the elements and, as a result, redefine the question being addressed. In other words, we recognize that inference using the usual methods of analysis will address households instead of individual adults or the mean of three serum cholesterol determinations instead of a single determination. This is the easier approach and should be considered if the sampling units define groups of elements that are relevant to the researcher.

After we have specified the elements and the sampling units, we are just about ready to obtain our sample (i.e., to identify which of the sampling units in the population will be included in the sample). To satisfy the universal assumption of statistics (that the sample be representative of the population) each sampling unit in the population must have a known nonzero probability of being included in the sample. To ensure that each sampling unit has a chance of inclusion in the sample, we need to identify and list all the sampling units in the population. The list from which the sample will be taken is called the *sampling frame* and the units that appear on the list are called *listing units*. Ideally, the listing units and the sampling units should be identical. In other words, each sampling unit should be represented in the sampling frame by one and only one listing unit. Failure of the sampling frame to include all the sampling units is likely to lead to a biased[3] sample about which statistical analysis can do little or nothing.

The next step in the process of sampling entails selecting a subset (ie, a sample) of the listing units and making observations or measurements on this subset. How this subset might be selected will be discussed in a moment, but all those methods assume that each listing unit that is selected will be observed or measured. Those that are observed or measured are called the *observational units*. Ideally the observational units are the same as the listing units that are selected to be in the sample. There are many common reasons that this might not be the case. For example, information might be lost, or an individual might refuse to participate in the study. Regardless of the reason, failure to make observations or measurements on all the listing units selected to be in the sample is likely to lead to a biased sample.[4]

Figure 2.1 summarizes the relationships among the various types of units.

---

[2] Other strategies for combining the information from repeated determinations also are used. For example, with three determinations, the highest and lowest values could be discarded. The same issues apply to these other strategies as to the mean of the determinations.

[3] Bias is a directional deviation from the truth. Imprecision, in contrast, is a nondirectional deviation from the truth, at least on the average.

[4] The only way that a biased sample can be avoided in this situation is if the listing units that are selected to be included in the sample but do not become observational units are identical, on the average, to the listing units that do become observational units.

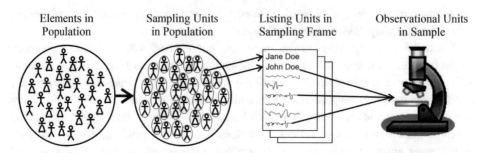

Fig. 2.1 The relationships among the units involved in taking a sample from a population. The elements are the units in the population for which estimation or inference is planned. Those elements are organized as sampling units each of which should appear as a listing unit on the sampling frame. Those listing units selected to be in the sample and contribute all data to the sample are called the observational units

Now that we understand some of the basic principles and terminology of sampling, we are ready to think about the specific mechanisms that can be used to draw a sample.

2.2   Model Sampling

In the process called *model sampling*, particular listing units are selected to become observational units by a researcher because that researcher believes that those listing units are representative of the sampling units in the population. One form of model sampling is *judgment sampling*, in which an expert (or group of experts) is asked to select the observational units. For example, suppose we wanted a sample of cities in the United States. To take a judgment sample, we would select cities we believed to be representative of all cities. Judgment sampling is the method that is most often used to select study sites. It is rarely used to select individuals to be included in a sample.

One type of model sampling that might be used to select individuals in a study is called *quota sampling*. In quota sampling, data collectors are instructed to make observations or measurements on a given number of representative units. It is this method of sampling that is used when interviewing the "man on the street."

To conduct model sampling, it is necessary to be able to recognize listing units that are representative of the sampling units in the population. This is a very difficult, if not impossible, task. For example, to choose a representative sample of cities, we might have to consider many, many different characteristics of cities and to differentiate between those characteristics that are of importance in each study. Seldom are we so familiar with the sampling units that we can make judgments about representativeness of listing units. More often, imperfect knowledge results in choices that create biased samples. Further, it is difficult to discern what is meant by a representative sample (e.g., how would you choose five cards that are representative of a deck of 52 cards?). The only time that model sampling is an advisable approach is when other methods are not practical.

2.3   **Probability Sampling**

An alternative to model sampling is an approach called *probability sampling*. The basic philosophy behind probability sampling is that chance, rather than knowledge, is used to select a representative sample. This reliance on chance circumvents the problem we encountered in model sampling of making incorrect choices because of imperfect knowledge or an inability to specify the meaning of a representative sample. When chance is used to select those listing units that will be observational units, then, on the average, the samples will contain observational units that are representative of the sampling units in the population.[5] Any particular sample, however, might be, by chance, distinctly unrepresentative of the population. This is the reason we use the statistical procedure of hypothesis testing to take the role of chance into account. This procedure helps us to address the probability of obtaining an unrepresentative sample when chance is used to select observational units.

In probability sampling, each listing unit has a known, greater-than-zero probability of becoming an observational unit. If each listing unit has the same probability of becoming an observational unit, we are using *simple random sampling*. It is necessary to use this method if we desire a sample in which the distribution of independent variable[6] values representative of the corresponding distribution in the population. In the next chapter, we will see that such a sample is needed for correlation analysis.

The most usual method for simple random sampling involves assigning a *random number* to each of the listing units. A random number is a number from a series of numbers that has been constructed without any apparent order. There are several sources of such random numbers. Many (older) statistical texts contain tables of random numbers. Today, a more convenient source of random numbers is the computer or pocket calculator. When random numbers are obtained from a computer or a pocket calculator, they are usually numbers between zero and one with each possible numeric value within that range having the same probability of occurring.

To decide which of the listing units will be selected as observational units, we first need to specify the proportion of the listing units we wish to include in the sample. This proportion is then used as the upper limit of the range of random numbers with possible values from zero to one that indicate a listing unit that will be selected as an observational

---

[5] This assumes all sampling units in the population occur as listing units in the sampling frame or that the listing units are, themselves, representative of the sampling units in the population.

[6] In statistical analysis, there are two types of variables. Independent variables represent the predictor data. Dependent variables represent the predicted (outcome) data. Types of variables are described further in Chap. 3.

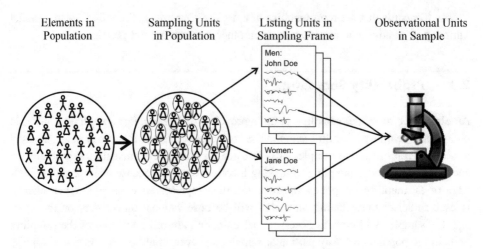

| Elements in Population | Sampling Units in Population | Listing Units in Sampling Frame | Observational Units in Sample |

Fig. 2.2 Schematic of the process of taking a stratified sample. In stratified sampling, listing units are separated into strata within each of which chance determines which of the listing units become observational units. Here, strata are defined by gender. The probability that a listing unit will become an observational unit can be different in different strata

unit.[7] For example, if we want to select 1% of the listing units, those listing units with random numbers between zero and 0.01 should be included among the observational units.

Simple random sampling is the most straightforward of the probability sampling methods. Samples selected with this method are the most straightforward to analyze. Often, however, there is good reason to use a more complicated probability sampling method. One way that the sampling method can be more complicated is by using different probabilities of becoming an observational unit for different groups of listing units. This method of sampling is called *stratified random sampling*.

In stratified sampling, the listing units are separated into groups (strata) specified by values of one or more characteristics. Within each of these strata, a random sample is taken so that a specific proportion of the listing units in a stratum become observational units. The proportions of listing units selected can be different for different strata. Thus, listing units in some strata can have a greater probability of being included in the sample than do listing units in other strata (Fig. 2.2).

Stratified sampling is used to ensure that groups of sampling units that might occur infrequently in the population are represented by a sufficient number of observational units in the sample to allow hypothesis testing with reasonable statistical power.[8] One type of study that always uses stratified sampling is a case-control study. In this type of

---

[7] This implies that each listing unit can be selected only once. In statistical terminology, we call this sampling without replacement. Nearly every sample in most research is taken without replacement.

[8] Recall from Chap. 1 that statistical power refers to the probability of avoiding a type II error in statistical inference. A type II error is accepting the null hypothesis when, in fact, it is false,

study, listing units are separated into two groups: cases (with the disease under study) and controls (without the disease under study). Then separate random selection processes for cases and controls are used to determine which listing units will become observational units. These selection processes are distinguished by a higher probability of any particular case becoming an observational unit than any particular control becoming an observational unit. Since case-control studies are performed only when cases are considerably less common than are controls, this stratified method of sampling is necessary to include a reasonable number of cases in the sample without requiring a very large total number of observational units.

Case-control studies are not the only type of study that use stratified sampling. In fact, whenever a researcher specifies the proportion of observational units in the sample that will have any characteristic, that researcher is using stratified sampling. Stratified sampling might be used to specify the number of individuals in a sample of various races, age groups, genders, etc. When the researcher selects the distribution of a characteristic of observational units in the sample, that researcher is using stratified sampling.

Intervention studies (such as a laboratory experiment or a randomized clinical trial) are an interesting special application of stratified sampling. Often, the participants in an intervention study are selected to be in the study with equal probability from listing units.[9] Thus, it might seem as if simple random sampling were used. The various interventions (such as treatments or doses), however, are assigned to the observational units in frequencies determined by the researcher. For example, half the participants might receive a new treatment and half might receive the standard treatment.[10] Even though assignment of the intervention occurs after observational units have been selected, the type of intervention an individual receives is thought of as a characteristic of the sampling units in the population. Since the researcher determines the proportion of the observational units that receive each intervention, the effect is the same as stratified sampling.

For a given sample's size, statistical power is greatest when there is the same number of observational units in each of the groups being compared.[11] An advantage of stratified sampling is that we can maximize statistical power by selecting the same number of observational units from each stratum. A disadvantage of using stratified sampling is that statistical analysis becomes a little bit more complex. For one thing, we must be certain to include in our analyses variables that represent the characteristics that were used to construct strata.

---

[9] The most frequently used method of obtaining participants in a randomized trial is taking all persons who qualify for entry into the study over a specified period of time. I will discuss this method of sampling a little later. Then, we will see that it is assumed that this method of sampling is the same as probability sampling.

[10] The assignment of treatments or doses is most often done using a process called randomization (thus, the term *randomized clinical trial*). This should not be confused with random sampling, a process of selecting persons to include in the sample.

[11] This will be proven in Chap. 5.

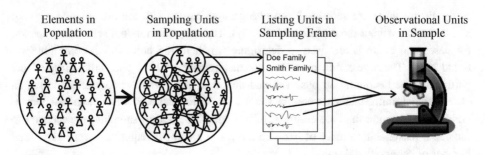

Elements in   Sampling Units   Listing Units in   Observational Units
Population   in Population   Sampling Frame   in Sample

Fig. 2.3 Schematic of a cluster sample in which sampling units (and, therefore, listing units) consist of more than one element

Earlier in this chapter I suggested that sampling units in the population can sometimes contain more than one element. For example, we might be interested in estimating the mean length of stay of hip fracture patients in the United States. Thus, the elements are persons admitted for hip fracture to any hospital anywhere in the United States. It would be very difficult for us to make a list of all of those patients even if we were willing to consider a very limited period of time. The way that we might approach this problem would be to obtain a list of hospitals rather than a list of patients and randomly select the hospitals for which we could examine the records of all hip fracture patients. We learned earlier that this strategy of sampling collections of elements rather than the elements themselves is known as cluster sampling (Fig. 2.3).

It is not unusual for research to rely on cluster sampling, but the use of this sampling method violates an assumption of commonly employed methods of statistical analysis. That assumption is that the observational units in a sample are statistically independent of one another. That is to say, these methods assume that data values for one individual in the sample are not influenced by the data values of other individuals. This is likely to be untrue when we sample clusters. For example, if we were to sample hospitals rather than individuals to estimate length of stay for hip fracture patients, we could expect to observe less variation among patients discharged from a given hospital than we would observe if we compared patients discharged from different hospitals. Thus, observation of length of stay for patients in any particular hospital are not statistically independent of one another.

The effect of violation of the assumption of independence among observational units is that a sample contains less information about the population than the sample's size suggests. This is because observational units that are not statistically independent produce data that are, to some degree, redundant. There are special statistical techniques that allow this redundancy to be taken into account, but most of these procedures are very complex. I will not describe these complex procedures in this text.

2.4   Other Methods of Sampling

When confronted with the task of sampling a certain proportion of a population, it is often tempting to use some regular pattern to identify those individuals who will be included in the sample. For example, if we wished to sample one-tenth of the patients seen at a particular clinic, we might decide to select every tenth patient in the order in which they present or to select all patients seen every other Tuesday for example. Sampling by using such a pattern is known as *systematic sampling*.

Systematic sampling is used in place of random sampling because it is easier to execute. It is, however, a dangerous method to select observational units. The danger is that listing units that fit into the pattern used to select observational units might be different than the listing units that do not fit into the pattern. For example, suppose different members of a family tend to be seen sequentially at the clinic from which we would like to draw a sample. If we selected every tenth listing unit from a sampling frame that listed patients in the order in which they were seem, there would be a very low probability of two members of the same family both becoming observational units. If what we are studying is in any way influenced by being in the same family, our sample would contain observational units that lack this influence. Thus, the sample would be biased.

Systematic sampling can be avoided. Even though taking a random sample is more complicated to perform than taking a systematic sample, the probability that there exists some unrecognized association between the pattern used to take a systematic sample and the relationships we are studying is too great to make the easier method advisable. For us to say that such an association does not exist is tantamount to taking a judgment sample. Rarely, if ever, do we understand relationships we are studying well enough to make this kind of judgment.

Another type of sampling is called *convenience sampling*. In this method of sampling, certain listing units are selected to become observational units because they are more easily included in the sample than are other listing units. This is the usual way that patients are recruited to participate in randomized clinical trials. In that case, all patients seen during a specific period of time at clinics participating in the study who meet the entrance criteria are asked to be in the study.[12]

Clearly, there is a potential for convenience sampling to result in a biased sample, and it seems to be a method that we should avoid. Unfortunately, nearly every sample that we take is, to some degree, a convenience sample. This is especially true when we think of time as one of the characteristics that is associated with the sampling units in the population. Generally, we would like to think of the results of our research to be applicable to individuals that exist in the population over an extensive period of time. In taking a sample, however, we are restricted to selection of sampling units as they exist

---

[12] Convenience sampling is the method used in randomized clinical trials that recruit newly diagnosed cases for whom a treatment is being investigated because of the virtual impossibility of constructing a sampling frame.

in the period of time that we can observe. Thus, we are taking a convenience sample of time. Since we cannot completely avoid taking convenience samples of some aspects of the sampling units in the population, we need to recognize how we are using this method of sampling and the danger it presents by resulting in a biased sample.

So far, we have seen several choices we can make among methods of obtaining a sample from a population. One choice we have is between model sampling (in which we use knowledge to select representative listing units) and probability sampling (in which we use chance to select representative listing units). Whenever practical, I recommend using probability sampling since statistical methods are designed to take this role of chance into account. Another choice is between elemental sampling (in which each sampling unit consists of only one element) and cluster sampling (in which sampling units consist of more than one element). I recommend using elemental sampling since the statistical procedures with which are familiar assume statistical independence among observational units. In addition to those choices, we can use systematic sampling (in which a specific pattern is used to select observational units) or convenience sampling (in which observational units are selected because they are more easily included in the sample). I recommend avoiding systematic sampling, but I recognize that it is often impossible to completely avoid convenience sampling.

Usually, the process that is used to select a sample consists of several of the methods we have described. For example, suppose we wished to estimate the prevalence of influenza among grade school students in a particular state and to make those estimates for students belonging to four different racial groups. First, we might select several school districts that we believe represent all the school districts in the state (judgment sampling). Then, we might randomly select several schools within each district (cluster sampling). Within each location, we might take a sample of students in each school in such a way that the four racial groups are represented by the same number of observational units in the sample (stratified sampling).

Such a process of sampling that involves more than one step is known as *multistage sampling*. Multistage sampling can make the process of drawing a sample easier than if sampling were to involve only one step. In the previous example, it would have been very difficult to take a stratified sample from a list of all students in the state. When multistage sampling is used, however, the method of statistical analysis must be designed to reflect the various sampling stages.

Now, let us look at an example to see how these various methods of sampling are combined in an actual research study.

Example 2.1 The following is a description of how the sample in the Framingham Heart Study was obtained.

In 1947, the Public Health Service (PHS) began to plan for an epidemiologic study of cardiovascular disease (CVD) with specific emphasis on arteriosclerotic and hypertensive cardiovascular disease. The plan was to obtain a sample of persons: (1) 30–59 years of age (to target those persons likely to develop CVD within a reasonable period of time), (2) free of overt CVD, and (3) from one geographic area (for reasons of cost). The maximum period of follow-up planned was 20 years.

Sample size estimates, based on incidence of CVD expected from previous surveys, suggested that 6,000 persons would need to be recruited for the planned study to allow observation of a sufficient number of cases of CVD within a 20-year period. Based on the age structure of the US population in 1947, it was predicted that a community of at least 25,000 would be required to yield 6.000 persons between 30 and 59 years of age. It was decided to set an upper limit of 50,000 on the community to be sampled to ensure that a community would be selected that would be small enough to maximize cooperation with the planned study.

In the middle of 1947, the Massachusetts State Health Commissioner asked the PHS to consider a number of Massachusetts communities as the site for the planned study. Among those considered, Framingham was chosen. Being 21 miles west of Boston, it was expected that Framingham offered access to epidemiologic and medical resources of Boston without losing its identity as a separate, stable community. In 1947, Framingham was an industrial and trade center with a population of 28,000. It had a town-meeting form of government which meant that the residents were used to working as a community on various projects. Further, Framingham was a successful site for the 1917–1923 community study of tuberculosis.

In October of 1947, the Heart Disease Epidemiology Study (HDES) began in Framingham to develop case-finding procedures for heart disease. In September of 1948 examinations were begun on volunteers. By July 1949, more than 1,500 volunteers had been examined and study facilities and staff had been established in Framingham. At this time, the HDES was transferred from the PHS Heart Disease Demonstration Section to the National Heart Institute (NHI).

In December 1949, NHI established a new sampling scheme (actually, reestablishing the original sampling objective). The 1950 town list (a publication based on an annual local census of persons 20 or more years of age) was stratified by family size and precinct of residence and arranged in order by address. Since there were approximately 10,000 residents of Framingham 30–59 years old in 1950, it was decided to take a two-thirds sample of households, recruiting all members of the household 30–59 years old. This sample was taken by selecting the first two households and skipping the third household according to street addresses.

The two-thirds sample of households identified 6,507 persons 30–59 years of age who were invited to participate in the study. Of those, 4,494 (68.8%) agreed. At the first exam-ination, it was discovered that 25 persons were not 30–59 years old, leaving a sample of 4,469. Because of the poor response to the first examination, it was decided to invite persons who had been volunteers for the PHS case-finding study (HDES) to participate in the NHI study. Of those volunteers, 740 were found to be eligible for entry into the study and agreed to participate. Thus, the original sample for the Framingham Heart Study was increased to 5,209 persons.

From this description, let us identify the elements, sampling units, sampling frame, listing units, observational units, and the method(s) of sampling used in the Framingham Heart Study.

To evaluate the sampling methods used in a study, we first need to determine what are the intended elements in the population to be sampled. In the Framingham Heart Study, the interest was to investigate factors associated with the development of cardiovascular disease. Thus, persons are the elements of interest.

This is a multistage sample. There are two main stages to the sampling process. The first stage involves selection of the site at which the study would be performed. This was done mostly as a convenience sample (various factors made a study in Framingham easier to perform). Although we are not told explicitly, we can suspect that the site was selected partly as a result of expert opinion that Framingham was, in some ways, a representative community (although it was distinctly unrepresentative as far as race is concerned). That is to say, the selection of Framingham probably reflects a degree of judgment sampling.

The second main stage of sampling involves selection of subjects from the Framingham community. The sampling frame was the 1950 town list. The listing units were households appearing in the town list. To prepare these for sampling (i.e., to establish the sampling units), these households were stratified by family size and precinct of residence and ordered by address. Then, two out of every three entries were selected. From this description, we can see that part of the sampling strategy involved cluster sampling. We know that this is the case since each listing unit (household) was the source of one or more elements (persons). Further, we see that stratified sampling was used. This is not as obvious since the proportion of households within each of the strata (defined by family size and precinct of residence) was the same as their proportion in the population. Even so, the sampling scheme guaranteed that this would be the case. That is different from simple random sampling in which chance would have determined the proportion from each stratum that actually occurred in the sample.

The mechanism of selecting the specific households from the sampling units to become observational units used a systematic, rather than random, process. That process resulted in every third house on each street being excluded from the sample. Perhaps this is not important, but systematic sampling was not needed, and its use created an unnecessary risk of some unforeseen bias.

The nature of the sample changed when it was found that too few subjects were recruited. At that point, volunteers were admitted. This lent a convenience sampling component to the sample. It would have been better if the original sampling frame were used to (randomly) select additional subjects.

Basic Statistical Methods

3

Abstract

This chapter describes common statistical methods that are used to generate the P-values needed for hypothesis testing. The chapter begins with a description of how a statistical method is selected to analyze a given set of data. Then, the methods are explained in a mostly nonmathematical manner, elucidating the hypotheses tested, how the tests can be performed using Microsoft® Office Excel®, and how to obtain the corresponding P-values. Finally, the P-values are interpreted relative to the null hypothesis. Numerous examples demonstrate these procedures.

To further examine hypothesis testing, we will look at the interpretation of a few basic statistical methods. As much as possible, we will avoid calculations. We will do this by using Microsoft® Office Excel®. To prepare to use Excel for statistical analysis, you must install an add-in that comes with Excel, but is inactive until the user installs it. To do that, start Excel and open a blank worksheet. Then, click on "File" in the main menu. Then, click on "Options" followed by "Add-ins." On the resulting screen, you will see "Manage" with a selection box. In that box, select "Excel Add-ins" and click "Go." If you are using Excel for Mac, in the file menu go to "Tools" then "Excel Addins." This will result in a menu being displayed that lists available add-ins. Put a check in the box next to "Analysis Toolpak" and click "Ok." Now click on "Data" in the main menu. Notice "Data Analysis" on the far right. Clicking on "Data Analysis" brings up a list of available statistical tests and other analysis tools.

© The Author(s), under exclusive license to Springer Nature Switzerland AG 2022
R. Hirsch, *Statistical Hypothesis Testing with Microsoft® Office Excel®*,
Synthesis Lectures on Mathematics & Statistics.
https://doi.org/10.1007/978-3-031-04202-7_3

3.1   Selecting a Test

To select a statistical test, we need to identify two things. The first of these is the type of variable that represents the data. The second is the type of data.

There are two types of variables: the dependent variable and the independent variable. The dependent variable represents the data of primary interest. It is what we designed the study to measure. If there is a causal relationship between the variables, the dependent variable represents what is caused. For example, suppose we want to look at the relationship between dietary sodium intake and diastolic blood pressure. Diastolic blood pressure is represented by the dependent variable. Every dataset has a dependent variable.

The independent variable represents the data that specifies the conditions under which we are interested in looking at the dependent variable. If there is a causal relationship, the independent variable represents the cause. In the example of a study looking at the relationship between dietary sodium intake and diastolic blood pressure, dietary sodium intake is represented by the independent variable. A data set can contain one, more than one, or no independent variables.

Data can be of two types: *continuous data* or *nominal data*.[1] Continuous data can be ordered, and the data values are evenly spaced. Dietary sodium intake and diastolic blood pressure are examples of continuous data. Continuous variables represent continuous data.

Nominal data are categories of things that cannot be ordered in a meaningful way. Nominal variables are dichotomous. Which of two treatments someone receives and whether someone is cured are examples of nominal data. Nominal variables represent nominal data.

The logical processes for selecting statistical tests are organized into flowcharts. Flowchart 3.1 shows the process for selecting a test to analyze a dataset that contains a continuous dependent variable.

Flowchart 3.2 shows the process for selecting a test to analyze a dataset that contains a nominal dependent variable.

Now, let us look at an example of using those flowcharts.

Example 3.1 Suppose we are interested in estimating the diastolic blood pressure related to the dietary sodium intake of persons in a population. To study this, we take a simple random sample from the population and determine both the daily average sodium intake and diastolic blood pressure for persons in the sample. What method should we use to analyze our data?

---

[1] Actually, there are three types of data. The third is ordinal data. Ordinal data can be ordered, but the spacing between values is undefined. We will not look at methods of analyzing ordinal data.

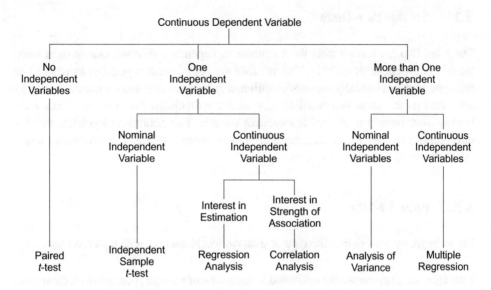

Flowchart 3.1 Selection of a statistical test for a continuous dependent variable

Flowchart 3.2 Selection of a statistical test for a nominal dependent variable

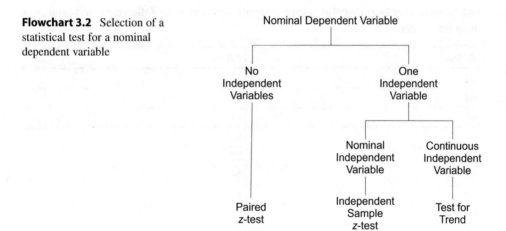

The dependent variable represents diastolic blood pressure (continuous data), so we use Flowchart 3.1. There is one independent variable, dietary sodium intake. That independent variable represents continuous data. We are told that our interest is in estimating diastolic blood pressure. This brings us to regression analysis.

3.2 Student's *t*-Tests

There are two Student's *t*-tests for a continuous dependent variable. One is used when there is no independent variable. This is called the *paired t-test*. It gets this name because this type of dataset usually consists of differences between two measurements of continuous data on the same individual or very similar individuals. The other Student's *t*-test is used when there is a nominal independent variable. This nominal independent variable distinguishes two groups of continuous dependent variable values. This Student's *t*-test is called the *independent sample t-test*.

3.2.1 Paired *t*-Test

Let us begin by considering the type of data we would analyze with a paired *t*-test.

Example 3.2 Suppose we are interested in the ability of an antihypertensive medication to lower diastolic blood pressure. To study this, we select a sample of 12 hypertensive persons and measure their diastolic blood pressure. Then we give them the medication for one week and measure their blood pressure again. Suppose we observe the following results and input them into Excel:

Before	After
91	79
94	84
96	84
93	81
99	85
95	85
92	81
90	78
86	76
97	85
98	86
88	78

Our interest in these data is to test the null hypothesis that the mean difference in diastolic blood pressure is equal to zero in the population versus the alternative hypothesis that it is not equal to zero. We use this two-tailed alternative hypothesis because it is possible that the medication could result in a higher diastolic blood pressure. We plan to set alpha to 0.05 to allow a 5% chance of making a type I error.

Although our interest is in the difference in diastolic blood pressure, we have set up the data in Excel as two columns of data. This is because the analysis tool for the paired *t*-test in Excel asks for two columns of data rather than the differences. We will see how that analysis tool works in the next example.

Example 3.3 To begin doing the paired *t*-test in Excel, access the statistical procedures by clicking "Data" in the main menu and then clicking "Data Analysis" in the submenu. This brings up the list of available analysis tools. From that list, select "t-Test: Paired Two-Sample for Means" then click "Ok." This invokes the following dialog box:

In this dialog box, the two columns of data are called "Variable 1" and "Variable 2." This is unfortunate because these columns are not two variables, but rather two observations of the same variable.

I have already identified the columns by selecting the "Variable 1 Range" and "Variable 2 Range." To do that, click the arrow on the right side of the input box, highlight the column of data, and click "Enter" on your keyboard. As part of the area I highlighted are the names of the columns in the first row, so I have put a check in the box next to "Labels." Clicking "Ok" results in the following output[2]:

t-test: paired two sample for means

	Before	*After*
Mean	93.25	81.83333333

---

[2] To make the entire output readable, you need to highlight it then click on "Home" in the main menu. In the submenu, click on "Format" and select "Autofit Column Width" from the dropdown box.

t-test: paired two sample for means		
Variance	16.20454545	11.78787879
Observations	12	12
Pearson correlation	0.957048729	
Hypothesized mean difference	0	
df	11	
t Stat	31.89105531	
P(T ≤ t) one-tail	1.71371E-12	
t Critical one-tail	1.795884819	
P(T ≤ t) two-tail	3.42743E-12	
t Critical two-tail	2.20098516	

The bottom line in that output is the two-tailed *P*-value. That is in the row labeled "P(T ≤ t) two-tail." It is equal to 3.43×10^{-12}. Since that is less than 0.05, we reject the null hypothesis and, through the process of elimination, accept the alternative hypothesis.

3.2.2   Independent Sample *t*-Test

Let us begin by considering the type of data we would analyze with the independent sample *t*-test.

Example 3.4 Imagine we are interested in comparing diastolic blood pressure between a group of persons who receive a new antihypertensive medication to a group of persons who receive the standard medication. Suppose we have 20 persons who are randomly assigned to receive one of the medications. After one week on the assigned medication, we measure diastolic blood pressure and observe the following results:

New	Standard
81	88
78	83
84	89
79	92
77	81
87	79
85	82
89	84

New	Standard
74	95
82	87

Our interest in these data is to test the null hypothesis that the difference in mean diastolic blood pressure is equal to zero in the population versus the alternative hypothesis that it is not equal to zero. We use this two-tailed alternative hypothesis because it is possible that the standard treatment could be better than the new treatment in lowering diastolic blood pressure. We plan to set alpha to 0.05 to allow a 5% chance of making a type I error.

In Student's *t-test*, a *t* statistic is calculated using the difference between the means, the variance of the data (how spread out the data values are), and the sample's size. Then a *P*-value is determined either from a table or from a computer program. Excel can find a two-tailed *P*-value for a Student's *t* statistic using the "T.DIST.2T" function or a one-tailed *P*-value using the "T.DIST.RT" function.[3] Excel also provides these *P*-values as part of its output from its Student's *t* procedure in Data Analysis. We will see this output in the next example.

There are two ways in which Student's *t-test* is done. The more common way is assuming the variances in the two groups are equal in the population from which the sample was taken. The other way is not assuming the variances are equal. The method for unequal variances often has a little less statistical power than the method for equal variances, so we use it only if we think the assumption has been violated. A good way to decide if the variances are unequal in the population is by comparing the variance estimates that are part of computer output from Student's *t*-test. We can be comfortable with the assumption of equal variances unless the variance estimates are substantially different.

In the next example, we will see how to perform Student's *t*-test in Excel and how to interpret the output.

Example 3.5 To begin doing Student's *t*-test in Excel, access the statistical procedures by clicking "Data" in the main menu and then clicking "Data Analysis" in the submenu. This brings up the list of available analysis tools. From that list, select "t-Test: Two-Sample Assuming Equal Variances" then click "Ok." This invokes the following dialog box:

---

[3] This is for a positive *t* statistic. If it is negative, the "T.DIST" function can be used to find the one-tailed *P*-value.

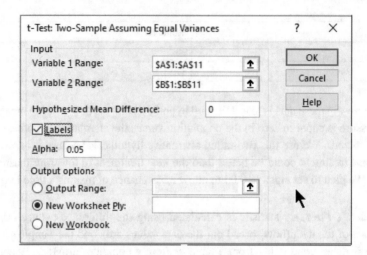

Again, Excel mislabels the two columns of data. They are not variable 1 and variable 2, but rather two groups of dependent variable values.

In this dialog box, I have already selected the "Variable 1 Range" and "Variable 2 Range." To do that, click the arrow on the right side of the input box, highlight the column of data, and click "Enter" on your keyboard as you did for the paired t-test dialog box. As part of the area I highlighted are the names of the groups in the first row, so I have put a check in the box next to "Labels." I have also put zero in the input box next to "Hypothesized Mean Difference" indicating the value in the null hypothesis, but this is not necessary since zero is the default value. Clicking "Ok" results in the following output:

t-test: two-sample assuming equal variances

	New	Standard
Mean	81.6	86
Variance	22.26666667	26
Observations	10	10
Pooled variance	24.13333333	
Hypothesized mean difference	0	
Df	18	
t Stat	− 2.002760526	
P(T ≤ t) one-tail	0.030249701	
t Critical one-tail	1.734063607	
P(T ≤ t) two-tail	0.060499402	
t Critical two-tail	2.10092204	

At the top of that output are the means and variances for the two groups. The first thing we do is compare the variance estimates. They are not substantially different, so Student's t-test assuming equal variances is the proper method to analyze these data. The next thing we look at is the two-tailed P-value. It is in the row labeled "P(T $\leq$ t) two-tail." That P-value is equal to 0.06. Since it is larger than 0.05, we fail to reject the null hypothesis.

If we had been able to say that it is impossible for the standard treatment to be better than the new treatment, we could have used a one-tailed alternative hypothesis and then the one-tailed P-value would have been appropriate. The one-tailed P-value is half of the two-tailed P-value. Thus, it is equal to 0.03. In this circumstance, we would have been able to reject the null hypothesis and, through the process of elimination, accepted the one-tailed alternative hypothesis.

3.3 Regression Analysis

The purpose of regression analysis is to estimate values of one continuous variable from values of another continuous variable. The next example shows the type of data for which we might perform a regression analysis.

Example 3.6 Suppose we are interested in how dose of a new antihypertensive medication effects the change in diastolic blood pressure. To study this, we measure diastolic blood pressure for 10 persons and then we give them each one of ten different doses of the medication. After a week on the medication, we measure diastolic blood pressure again. We subtract the second measurement from the first measurement to get the change in diastolic blood. Suppose we get the following when we enter the results in Excel:

Dose	Change
0.1	8
0.2	4
0.3	10
0.4	8
0.5	15
0.6	12
0.7	14
0.8	18
0.9	15
1	20

Our interest in those data is to estimate the change in diastolic blood pressure associated with the dose of medication. Hypothesis tests include tests for the parameters of the regression equation (described shortly) and a test of the null hypothesis that knowing dose does not help estimate the change in diastolic blood pressure. We will perform those hypothesis tests with a two-tailed alternative hypothesis and an alpha of 0.05, allowing a 5% chance of making a type I error.

To begin a regression analysis, it is helpful to examine the data graphically. We do this with a *scatter plot*. A scatter plot has the independent variable on the horizontal (X) axis and the dependent variable on the vertical (Y) axis. We can generate a scatter plot in Excel by clicking on "Insert" in the main menu and then "Scatter" in the list of charts in the secondary menu. Then right click the plot area and pick "Select Data" from the dropdown menu. The "X values" are the independent variable values and the "Y values" are the dependent variable values. Axis labels can be added to the scatter plot by clicking on the scatterplot and then "Chart Design" in the main menu. Selecting "Add Chart Element" from the submenu allows you to add things such as axis labels to the plot. The next example shows a scatter plot for the data in Example 3.6.

Example 3.7 Let us use Excel to generate a scatter plot for the data in Example 3.6.
We get the following scatter plot in Excel:

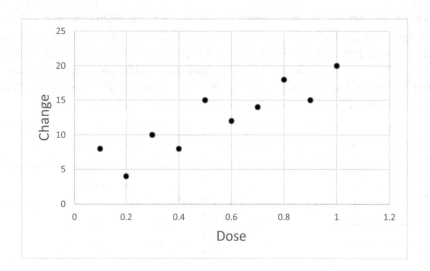

To estimate dependent variable values from independent variable values, we need to fit a line to the points in the scatter plot. The simplest line is a straight line. A straight line has two parameters: the *slope* and the *intercept*. The slope tells us how much the dependent variable value changes as the independent variable is increased by one unit. The intercept tells us the value of the dependent variable when the independent variable is equal to zero in the population. In statistics, the slope is symbolized with the letter "b" and the intercept is symbolized with the letter "a." Then, the formula for a straight line is as shown in Eq. 3.1.

$$\hat{Y} = a + b \cdot X \tag{3.1}$$

The "hat" over the "Y" in Eq. 3.4 is the statistical way to symbolize an estimated value.

To estimate the slope and the intercept, we use regression analysis. To perform a regression analysis in Excel, we select "Data" from the main menu and click on "Data Analysis" in the submenu. Then we scroll down and select "Regression" in the popup menu. This generates a dialog box shown in Example 3.8.

Example 3.8 Selecting the "Regression" analysis tool. results in the follow dialog box.

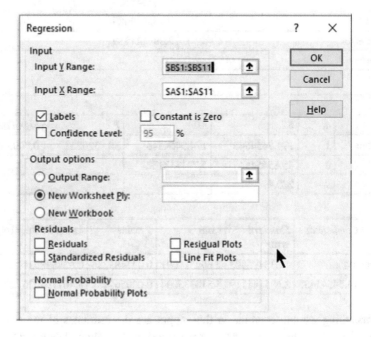

In this dialog box, I have already selected the "Input Y Range" and "Input X Range." To do that, click the arrow on the right side of the input box, highlight the column of data, and click "Enter" on your keyboard. As part of the area I highlighted are the names of the groups in the first row, so I have put a check in the box next to "Labels." Clicking "Ok" results in the regression output.

We will look at the regression output in the next example, but first, we need to discuss the hypothesis tests associated with regression analysis. There are three hypothesis tests. One tests the null hypothesis that the slope is equal to zero in the population and another tests the null hypothesis that the intercept is equal to zero in the population. The third tests what is called the *omnibus null hypothesis*. The omnibus null hypothesis states that knowing the value of the independent variable does not help estimate the value of the dependent variable.

Example 3.9 Let us interpret the regression output for the data in Example 3.6.
Excel provides us with the following output:

SUMMARY OUTPUT

Regression statistics	
Multiple R	0.889914897
R^2	0.791948523
Adjusted R^2	0.765942089
Standard error	2.394121589
Observations	10

ANOVA

	df	SS	MS	F	Significance F
Regression	1	174.5454545	174.5454545	30.4520222	0.000561481
Residual	8	45.85454545	5.731818182		
Total	9	220.4			

	Coefficients	Standard error	t stat	P-value	Lower 95%	Upper 95%
Intercept	4.4	1.635496403	2.690314691	0.027483234	0.628538531	8.171461469
Dose	14.54545455	2.635841119	5.518335094	0.000561481	8.467194026	20.62371506

The first thing we can examine in this output are the estimates of the slope and the intercept. These are in the bottom table of the output in the column labeled "Coefficients." The estimate of the intercept is in the row labeled "Intercept" and the estimate of the slope is in the row labeled with the name of the independent variable.[4] From those estimates, we know that the regression equation is:

---

[4] The reason the row with the slope is labeled with the name of the independent variable is the fact that regression analysis can include more than one independent variable. We will see that when we discuss multiple regression analysis.

$$\widehat{\text{Change}} = 4.4 + 14.55 \cdot \text{Dose}$$

To estimate the change in diastolic blood pressure, we plug a value for dose in that equation. For a dose of 0.5, we estimate a change in diastolic blood pressure of 11.7 mm Hg.

The P-values testing the null hypotheses that the parameters of the regression line are equal to zero in the population are in the column labeled "P-value." The P-value for the intercept is 0.027. Since this is less than 0.05, we can reject the null hypothesis that the intercept is equal to zero in the population and, through the process of elimination, accept the alternative hypothesis that the intercept is not equal to zero. The P-value for the slope is equal to 0.00056. Since this is less than 0.05 we can reject the null hypothesis that the slope is equal to zero in the population and, though the process of elimination, accept the alternative hypothesis that the slope is not equal to zero.

The P-value testing the omnibus null hypothesis in is the middle table titled "ANOVA" It is in the first row of the table under the label "Significance F." In this output, the P-value for the omnibus null hypothesis is 0.00056. Since that is less than 0.05, we reject the omnibus null hypothesis that knowing dose does not help estimate the change in diastolic blood pressure and accept the alternative hypothesis that knowing dose does help estimate the change in diastolic blood pressure.

Note that the P-value for the omnibus null hypothesis is the same as the P-value for the null hypothesis that the slope is equal to zero in the population. This will always be the case for regression analyses that include one independent variable. This is because a slope of zero is a horizontal line and we cannot estimate different values of the dependent variable corresponding to different values of the independent variable from a horizontal line. Thus, the omnibus hypothesis is true if and only if the null hypothesis that the slope is zero in the population is true.

3.4   Correlation Analysis

Correlation analysis is performed on the same type of data sets that use regression analysis. However, these two analyses address different questions. Regression analysis is concerned with estimation of dependent variable values from values of the independent variable. Correlation analysis is concerned with the strength of the association between two variables.

By strength of an association, we mean the consistency in direction and magnitude with which the value of one variable changes with an increase in the value of the other variable. If the first variable tends to increase in value as the other variable increases, we

Fig. 3.1 Perfect direct
association with a correlation
coefficient equal to 1.0

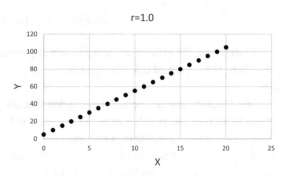

Fig. 3.2 Strong direct
association with a correlation
coefficient equal to 0.8

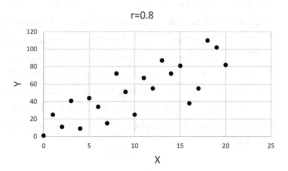

say that they have a *direct association*.[5] If the first variable tends to decrease in value as
the other variable increases, we say that they have an *inverse association*.[6]

The strength of an association between two continuous variables is reflected by the
value of the *correlation coefficient*. We symbolize correlation coefficients with the letter r.
Correlation coefficients have values ranging from minus one to positive one. A correlation
coefficient of positive one indicates a perfect direct association. A correlation coefficient
of negative one indicates a perfect inverse association. A correlation coefficient of zero
indicates no association.

To visualize the strength of association indicted by values of the correlation coefficient,
examine Figs. 3.1, 3.2, 3.3 and 3.4. Figure 3.1 illustrates a perfect direct association with
a correlation coefficient equal to one. Figure 3.2 illustrates a strong direct association with
a correlation coefficient equal to 0.8. Figure 3.3 illustrates a weak direct association with a
correlation coefficient equal to 0.2. Figure 3.4 illustrates no association with a correlation
coefficient equal to 0.0.

Although a correlation coefficient can be estimated whenever we have a data set that
includes two continuous variables, it is not always appropriate. The value of the correlation coefficient is influenced by how the independent variable is sampled. If extreme

---

[5] In regression analysis, a direct association is identified by having a positive slope.

[6] In regression analysis, an inverse association is identified by having a negative slope.

Fig. 3.3 Weak direct association with a correlation coefficient equal to 0.2

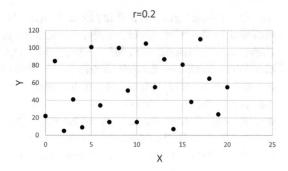

Fig. 3.4 No association with a correlation coefficient equal to 0.0

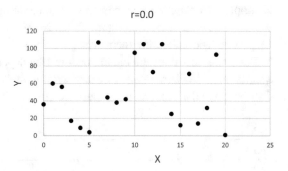

values are selected the correlation coefficient will be closer to one than if moderate values are selected. For example, consider the data in Fig. 3.2. The independent variable (X) has been selected uniformly throughout its range. If we were to consider only the ten most extreme values (0–4 and 16–20), the correlation coefficient is closer to one (0.85). If we were to select the eleven middle values (5–15), the correlation coefficient is further from one (0.69).

Since the value of the correlation coefficient is influenced by how the independent variable is sampled, the way it is sampled should be so that its distribution in the sample reflects its distribution in the population if we want the estimate to reflect the strength of association in the population. A sample in which the distribution of the independent variable in the sample is representative of its distribution in the population is called a *naturalistic sample.* A naturalistic sample results from taking a simple random sample.

A sample for which its distribution is determined by the researcher is called a *purposive sample.* A purposive sample results from taking a stratified random sample. Correlation analysis is appropriate only if we have a naturalistic sample.[7]

The null hypothesis in correlation analysis is that the correlation coefficient is equal to zero in the population. A *P*-value for that null hypothesis is not part of the output from

---

[7] Regression analysis, on the other hand, is appropriate regardless of how the independent variable is sampled.

the "Correlation" analysis tool in "Data Analysis" in Excel. Rather, the only information provided by the "Correlation" analysis tool is the estimate of the correlation coefficient. Fortunately, the correlation coefficient and the P-value testing the null hypothesis that the correlation coefficient is equal to zero in the population versus the alternative hypothesis that it is not equal to zero are part of the output from the "Regression" analysis tool. The estimate of the correlation coefficient is labeled "Multiple R" in the regression analysis output.[8] The P-value for the test of the null hypothesis that the correlation coefficient is equal to zero in the population is the same as the P-value testing the omnibus null hypothesis. This is illustrated in the next example.

Example 3.10 Suppose we are interested in the strength of the association between dietary sodium intake and blood pressure. To study this, we select a simple random sample of 15 persons and determine their average daily sodium intake (the independent variable) and measure their diastolic blood pressure (the dependent variable). This is a naturalistic sample since the distribution of dietary sodium intake in the sample is representative of its distribution in the population. Suppose we observe the following results:

NA	DBP
1.5	70
2.3	73
2.8	80
1.9	78
3.4	83
2.6	72
5.1	91
2.2	78
2.1	82
3.8	79
2	76
2.3	77
2.1	75
2.8	76
1.7	78

---

[8] It gets that label because the "Regression" analysis tool can perform multiple regression analysis with more than one independent variable and then the correlation coefficient reflects the correlation of the entire collection of independent variables and the dependent variable.

Let us use the "Regression" analysis tool to perform a correlation analysis with an alpha of 0.05.

SUMMARY OUTPUT

Regression statistics	
Multiple R	0.743675201
R^2	0.553052804
Adjusted R^2	0.518672251
Standard error	3.507002996
Observations	15

ANOVA

	df	SS	MS	F	*Significance F*
Regression	1	197.8454232	197.8454232	16.08621001	0.001481755
Residual	13	159.8879102	12.29907001		
Total	14	357.7333333			

	Coefficients	*Standard error*	*t stat*	*P-value*	*Lower 95%*	*Upper 95%*
Intercept	67.46509579	2.746951186	24.55999078	2.80611E-12	61.53066855	73.39952304
NA	4.04206122	1.007803865	4.010761775	0.001481755	1.864833337	6.219289103

In the first table, the estimate of the correlation coefficient is to the right of the label "Multiple R." That correlation coefficient is equal to 0.74. The *P*-value for the omnibus null hypothesis is in the second table below the label "Significance F." That is also the *P*-value testing the null hypothesis that the correlation coefficient is equal to zero in the population. That *P*-value is equal to 0.0015. Since that *P*-value is less than 0.05, we reject the null hypothesis and, through the process of elimination, accept the alternative hypothesis that it is not equal to zero.

To interpret the numeric magnitude of the correlation coefficient, we often take its square. The correlation coefficient squared appears in the regression output in the row of the first table labeled "R^2." The square of the correlation coefficient tells us the proportion of the variation in the dependent variable is associated with variation in the independent variable. In Example 3.10, the value of the square of the correlation coefficient is 0.55. That implies that 55% of the variation in diastolic blood pressure is associated with dietary sodium intake.

3.5   Analysis of Variance

One nominal independent variable separates dependent variable values into two groups. When we have more than one nominal independent variable, the dependent variable values are divided into more than two groups.[9] In fact, k groups are specified by $k - 1$ nominal independent variables. When the dependent variable represents continuous data and we have more than one nominal independent the method we use to analyze those data is called *analysis of variance*. Analysis of variance (or *ANOVA*) compares the means of all the groups, testing the null hypothesis that the means are all equal to the same value in the population.[10] The next example shows the type of dataset that we would analyze with ANOVA.

Example 3.11 Suppose we have three drugs that are designed to lower serum cholesterol. To compare them to each other and to a low fat diet we take a sample of 40 persons and randomly assign each to receive one of the drugs or to follow a low-fat diet for a period of four weeks. Then, we measure each person's serum cholesterol. Suppose we observe the following results:

Drug A	Drug B	Drug C	Diet
129	162	141	188
182	173	146	209
175	155	138	174
174	149	135	183
184	161	159	192
136	160	166	177
125	165	185	162
142	155	172	155
142	169	155	185
131	151	163	175

Our interest in these data is to compare their means. With ANOVA we test the omnibus null hypothesis that all four means are equal to the same value in the population. The alternative hypothesis is that not all the means are equal to the same value. Let us test that null hypothesis with an alpha of 0.05, allowing us a 5% chance of making a type I error.

---

[9] Nominal variables are always dichotomous.

[10] Analysis of variance might seem a strange name for an analysis that compares means. It gets this name from the way it compares means. ANOVA considers two sources of variation: the variation within groups and the variation between groups. If the variation between groups is much greater than the variation within groups, the means of those groups must be different.

We perform an ANOVA in Excel using the ANOVA analysis tool. If we look at the list of analysis tools in "Data Analysis," we see there are three ANOVA analysis tools. One thing that separates them is whether they have one or two *factors*. A factor is a characteristic. The data in Example 3.11 have one characteristic or factor: treatment. That factor has four categories. If we also wanted to pay attention to gender, we could add that as another factor with two categories. Then, we would divide the persons into eight groups and use the "Anova: Two-Factor With Replication" analysis tool. There is replication because there would be more than one person in each group. The "Anova: Two-Factor Without Replication" is used when there is only one person in each group.

ANOVA with more than one factor is called a *factorial* ANOVA. ANOVA with only one factor is called a *one-way* ANOVA. Since our data have categories of only one factor, it is a one-way ANOVA and we use the "Anova Single Factor" analysis tool. This is illustrated in the next example.

Example 3.12 To perform a one-way ANOVA, with click on "Data" in the main menu bar and then select "Data Analysis" from the submenu. That invokes a popup menu that lists the analysis tools. From that menu, we select "Anova: Single Factor." That opens the following dialog box.

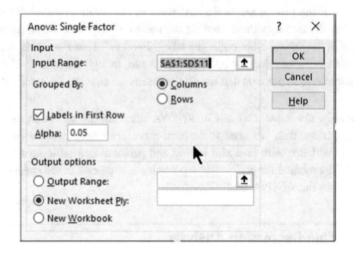

In that dialog box, you click on the arrow in the input box labeled "Input Range:" and highlight all the data. Make sure the circle by Columns is selected. I included the data names in the area I highlighted, so I checked the box next to "Labels in First Row." Clicking on "Ok" results in the following output:

Anova: single factor

SUMMARY

Groups	Count	Sum	Average	Variance
Drug A	10	1520	152	565.7777778
Drug B	10	1600	160	59.11111111
Drug C	10	1560	156	260.6666667
Diet	10	1800	180	233.5555556

ANOVA

Source of variation	SS	df	MS	F	P-value	F crit
Between groups	4640	3	1546.666667	5.528196982	0.003157921	2.866265551
Within groups	10072	36	279.7777778			
Total	14712	39				

The means of the groups are in the first table in the column labeled "Average." The *P*-value testing the null hypothesis that all the means are equal to the same value in the population is in the second table under the label "*P*-value." That *P*-value is 0.0032. Since that is less than 0.05, we reject the null hypothesis and, though the process of elimination, accept the alternative hypothesis that not all the means are equal to the same value.

Unfortunately, the hypothesis test in ANOVA does not tell us which means are different. To determine that, we need to perform a *posterior test*. Excel does not have an analysis tool that helps with posterior testing and performing a posterior test by hand is complicated. The method for performing a posterior test appears in the appendix for those who wish to take the ANOVA analysis further.

3.6   Multiple Regression Analysis

When we have a dataset that includes a continuous dependent variable and more than one continuous independent variable, we need to perform a *multiple regression analysis*. A common reason for having more than one continuous independent variable is that there is a desire to control for one or more continuous variables when examining the relationship between a particular independent variable and the dependent variable. Including variables for which you want to control in the regression analysis controls for their effect. To see how this works, let us look at an example.

Example 3.13 Suppose we have developed a new exercise program designed to help persons lose weight. We are interested in estimating the amount of weight lost in a three-month period based on the average time spent exercising per week. When looking at this relationship, we want to control for the average calories eaten per day. To study this, we measure each person's weight before and after the three -month period and calculate the difference (before-after). We also ask each person to record the time spent exercising each day as well as keep a food log recording everything they eat each day. Suppose we make the following observations and input them into Excel.

Change	Time	Calories
6	7	3599
−5	0	4328
2	1	2407
11	4.5	2913
4	19	3301
−2	1	3652
16	17	1800
4	10	2968
−5	15.5	5143
9	9.5	3788
19	24.5	1900
19	8.5	1925
17	20.5	2446
16	18	1438
6	4	1857
11	6.5	2476
11	15.5	1850
4	5	3027
9	22.5	3314
6	14	3387

To analyze data such as these with multiple regression analysis, we use the same "Regression" analysis tool we previously used for regression analysis with one independent variable. This is illustrated in the next example.

Example 3.14 To perform a multiple regression analysis, we click on "Data" in the main menu and select "Data Analysis" in the submenu. From the dropdown menu, we select the "Regression" analysis tool. this invokes the following dialog box:

In that dialog box, we select the dependent variable values in the input box labeled "Input Y Range" and we select all the independent variable values in the input box labeled "Input X Range." If the variable names are part of the areas highlighted, we check the box labeled "Labels." Clicking on "Ok" results in the following output:

SUMMARY OUTPUT

Regression statistics

Multiple R	0.860917087
R^2	0.741178231
Adjusted R^2	0.710728612
Standard error	3.928729094
Observations	20

ANOVA

	df	SS	MS	F	Significance F
Regression	2	751.406491	375.7032455	24.34113251	1.02449E-05
Residual	17	262.393509	15.43491229		
Total	19	1013.8			

	Coefficients	Standard error	t stat	P-value	Lower 95%	Upper 95%
Intercept	19.1270422	3.459683245	5.528553004	3.68359E-05	11.82774859	26.4263358
Time	0.356922784	0.12179911	2.930421934	0.009339948	0.099949124	0.613896444
Calories	− 0.00529065	0.00096118	-5.50427983	3.8699E-05	−0.00731858	−0.00326272

Our interest in performing this regression analysis was to estimate the change in weight from the time spent exercising. To do that, we need the multiple regression equation. The multiple regression equation is similar to the regression equation for one independent variable (Eq. 3.1) with the additional independent variable added to the equation as follows:

$$\widehat{Change} = 19.127 + 0.357 \cdot Time - 0.005 \cdot Calories$$

For example, we could use that equation to estimate the change in weight we could expect for someone who averaged 10 h per week exercising. Since we are controlling for the average calories per day consumed, we need to specify a value for the number of calories in our estimate. Suppose we were to select 2,000 cal. Then, plugging those values in the regression equation, we get the change in weight we would expect to observe, on the average. That change in weight is 12.1 pounds.

There are four hypothesis tests in this multiple regression output. The first P-value is in the second table under the label "Significance F." That P-value is 1.024×10^{-5}. It tests the omnibus null hypothesis that knowing Time and Calories does not help estimate the change in weight. Since that P-value is less than 0.05, we can reject the omnibus null hypothesis and, though the process of elimination, accept the alternative hypothesis that knowing the values of the independent variables does help estimate dependent variable values.

Other null hypotheses tested in multiple regression are that the intercept and the slopes (called *regression coefficients* in multiple regression) are equal to zero in the population. The P-values for those null hypotheses are in the third table in the column labeled "P-value." The P-value for the null hypothesis that the intercept is equal to zero is 3.684×10^{-5}. Since that is less than 0.05, we can reject the null hypothesis that the intercept is equal to zero. The P-value for the null hypothesis that the regression coefficient for Time is equal to zero in the population is equal to 0.00934. Since that is less than 0.05, we can reject the null hypothesis that the regression coefficient for Time is equal to zero in the population. The P-value testing the null hypothesis that the regression coefficient for Calories is equal to zero in the population is 3.870×10^{-5}. Since that is less than 0.05, we can reject the null hypothesis that the regression coefficient for Calories is equal to zero in the population.

3.7   Z-Tests for Nominal Data

For all the analyses for continuous dependent variables, we were able to use an analysis tool. There are no analysis tools for a nominal dependent variable. This means we must do these analyses manually. Fortunately, the calculations are not too complicated.

3.7.1   Paired z-Test

When we have a nominal dependent variable and no independent variables, we use something called a *paired z-test*. It has this name because this type of dataset usually results from a study with a paired design. This is where the same person makes a selection between two things. This is often called a *preference study*. The next example describes a preference study.

Example 3.15 Suppose we are interested in comparing two treatments for migraine relief: a new medication and the current leading treatment. To study this, we take a sample of 100 persons who have an average of one migraine headache per week. We give them both medications in a blinded manner and ask them to use each for one month each time they have a migraine. Then we ask them which of the two medications they prefer. Suppose 60 persons say they preferred the new medication.

Our interest in these data is to test the null hypothesis that there is no preference in the population. That is saying that the proportion favoring either group will be 0.5. The alternative hypothesis is that there is a preference. We use this two-tailed alternative hypothesis because it is possible that there could be a preference for either treatment. In testing the null hypothesis, we will allow a 5% chance of making a type I error by comparing the P-value to 0.05.

The null hypothesis in a preference study is that there is no preference between the two things offered in the population. That is to say, 50% of the persons will prefer each of the offerings. To test that null hypothesis, we calculate a *standard normal deviate* or *z-value*. That calculation is shown in Eq. 3.2.

$$z = \frac{p - \theta}{\sqrt{\frac{\theta \cdot (1-\theta)}{n}}} \tag{3.2}$$

where

p observed proportion
θ proportion in the null hypothesis (0.5)
n sample's size.

To interpret a z-value, we use Excel to determine the corresponding P-value. We do that by using an Excel function. That function is "NORM.S.DIST." Eq. 3.3 shows that function for a positive z-value.

$$= 2 * (1 - \text{NORM.S.DIST}(z, \text{TRUE})) \qquad (3.3)$$

Equation 3.4 shows that function for a negative z-value.

$$= 2 * \text{NORM.S.DIST}(z, \text{TRUE}) \qquad (3.4)$$

The next example shows the test of the null hypothesis that there is no preference in the population for the study in Example 3.15.

Example 3.16 In Example 3.15, we are told that 60 of 100 persons preferred the new treatment. That corresponds to a proportion of 0.6. The following calculation results in a z-value that represents a proportion of 0.6. Using Eq. 3.2, we get:

$$z = \frac{0.6 - 0.5}{\sqrt{\frac{0.5 \cdot (1 - 0.5)}{100}}} = 2.00$$

To interpret that z-value, we use the function in Eq. 3.3. The resulting P-value is 0.0455. Since that P-value is less than 0.05, we can reject the null hypothesis that there is no preference in the population and, through the process of elimination, accept the alternative hypothesis that there is a preference.

3.7.2 Independent Sample z-Test

When we have a nominal dependent variable and one nominal independent variable, we use an *independent sample z-test*. Let us begin by considering the type of data for which we would use an independent sample z-test.

Example 3.17 Suppose we have a new antiviral medication for treatment of patients with early covid-19 infections. Our interest is in how well the medication does at keeping patients from being hospitalized. To examine this, we identify 1,000 newly diagnosed covid-19 patients and randomly assign 500 to receive the new medication and 500 to receive the current antiviral medication. Then we follow each patient for the course of their infection and determine how many are hospitalized. Among the patients receiving the standard treatment, 100 (20%) are hospitalized. Among the 500 receiving the new treatment, 70 (14%) are hospitalized.

Our interest in these data is to test the null hypothesis that the difference in proportions of persons being hospitalized is equal to zero in the population versus the alternative that it is not equal to zero. We choose a two-tailed alternative hypothesis because it is possible that the new treatment is worse than the standard treatment. To test this null hypothesis, we select an alpha of 0.05 giving us a 5% chance of making a type I error.

Data such as these are often organized in a 2×2 *table*. A 2×2 table consists of four cells resulting from the intersection of two columns and two rows. The two columns are identified by the two groups being compared and the two rows are identified by the two possible outcomes (having or not having the event of interest).[11] The next example organizes the data from Example 3.3 in a 2×2 table.

Example 3.18 Let us organize the data in Example 3.17 in a 2×2 table.

		Groups		
		Standard	New	
Hospitalized	Yes	100	70	170
	No	400	430	830
		500	500	1,000

The analysis of 2×2 table data involves calculation of a standard normal deviate or z-value. Equation 3.5 shows that calculation.[12]

$$z = \frac{p_1 - p_2}{\sqrt{\overline{p} \cdot (1 - \overline{p}) \cdot \left(\frac{1}{n_1} + \frac{1}{n_2}\right)}} \tag{3.5}$$

where

p_1 proportion of observations in group 1 with the event
p_2 proportion of observations in group 2 with the event
$\overline{p}$ marginal proportion of observations with the event
n_1 number of observations in group 1
n_2 number of observations in group 2.

The next example shows the calculation of the z-value in an independent sample z-test.

Example 3.19 Let us calculate the z-value for the data in the 2×2 table in Example 3.18.

---

[11] There is not a standard orientation of 2×2 tables. Sometimes the rows specify the two groups, and the columns specify the events.
[12] 2×2 table data are often analyzed using a chi-square test, The z-test gives exactly the same P-value as the most commonly used chi-square test (Pearson's chi-square).

From the 2×2 table in Example 3.18 we can determine the proportion of persons in the standard medication group who are hospitalized ($100/500 = 0.20$) and the proportion of persons in the new medication group who are hospitalized ($70/500 = 0.14$). The marginal proportion is determined from the values to the right of the 2×2 table ($170/1,000 = 0.17$). The number of observations in the two groups are the values below the 2×2 table (500). With those values, we are ready to use Eq. 3.5.

$$z = \frac{0.20 - 0.14}{\sqrt{0.17 \cdot (1 - 0.17) \cdot (1/500 + 1/500)}} = 2.526$$

To find the P-value associated with that z-value, we use the Excel function "NORM.S.DIST" as shown in Eqs. 3.3 and 3.4.

The next example determines and interprets the P-value for the z-value calculated in Example 3.19.

Example 3.20 Let us determine and interpret the P-value for the result of Example 3.19

Using the function in Eq. 3.3 in Excel, we get a P-value of 0.012. Since this is less than 0.05, we reject the null hypothesis and, through the process of elimination, accept the alternative hypothesis that the difference in the proportions hospitalized is not equal to zero in the population.

3.7.3 Test for Trend

When we have a nominal dependent variable and a continuous independent variable, we perform a regression analysis called a *test for trend*. Like regression analysis, the objective for the test for trend is to estimate the probability of the event represented by the dependent variable for values of the independent variable. To do this, we estimate parameters of a regression equation. The equation is shown in Eq. 3.6.

$$\hat{p} = a + b \cdot X \tag{3.6}$$

where

$\hat{p}$ estimated probability of the dependent variable event
a intercept
b slope
X value of the independent variable.

Equation 3.6 for a nominal dependent variable is very similar to the regression equation for a continuous dependent variable (Eq. 3.1). The only difference is what is on the lefthand side of the equal sign.

To estimate the parameters of that regression equation, we can use the "Regression" analysis tool in Excel. To do that, we need to represent the dependent variable event quantitatively. We do that by using a *dummy variable*.[13] A dummy variable has the values of zero and one. One signifies that the event occurred and zero signifies that the event did not occur. Example 3.21 shows a dataset that includes a dummy variable.

Example 3.21 Suppose we have a new treatment for bronchitis, and we are interested in its effectiveness at various doses. To study this, we randomly assign 20 persons with newly diagnosed bronchitis one of four different doses. Suppose we observe the following results:

Dose	Cure	Dummy
5	No	0
5	Yes	1
5	No	0
5	No	0
5	No	0
10	No	0
10	Yes	1
10	No	0
10	No	0
10	Yes	1
15	Yes	1
15	No	0
15	Yes	1
15	No	0
15	Yes	1
20	Yes	1
20	Yes	1
20	Yes	1
20	No	0
20	Yes	1

---

[13] This is also called an *indicator variable*.

In that dataset, I have created a dummy variable to represent the nominal dependent variable (Cure) that is equal to one if the person was cured and equal to zero if the person was not cured.

Once we have a dataset that includes a dummy variable representing the nominal dependent variable, we can use the "Regression" analysis tool in Excel to estimate the slope and intercept of the regression equation that we can use to estimate the probability of being cured for various doses. In that regression analysis, we set the dummy variable to be the dependent variable. This is illustrated in the next example.

Example 3.22 For the data in Example 3.21 let us estimate the slope and intercept of the regression equation for being cured as the dependent variable and dose as the independent variable. To do that, we click on "Data" in the main menu and select "Data Analysis" from the submenu. This creates a popup menu with a list of the analysis tools available in Excel. From that list, we select the "Regression" analysis tool. That selection results in the following dialog box.

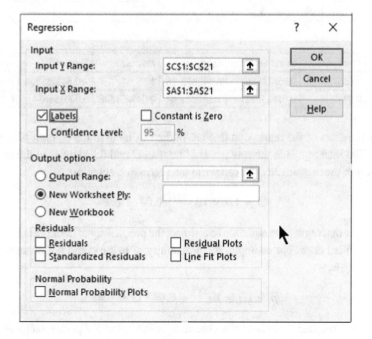

In that dialog box, we identify the dummy variable by clicking the arrow in the input box labeled "Input Y Range," highlighting the dummy variable, and clicking "Enter" on the keyboard. Then, we identify the independent variable by clicking the arrow in the input box labeled "Input X Range" and highlighting the column with the doses. If we have included the variables names in the areas we highlighted, we check the box labeled "Labels." Clicking "Go" results in the following output:

SUMMARY OUTPUT	
Regression statistics	
Multiple R	0.447213595
R^2	0.2
Adjusted R^2	0.155555556
Standard error	0.471404521
Observations	20

ANOVA					
	df	*SS*	*MS*	*F*	*Significance F*
Regression	1	1	1	4.5	0.048037528
Residual	18	4	0.222222222		
Total	19	5			

	Coefficients	*Standard error*	*t stat*	*P-value*	*Lower 95%*	*Upper 95%*
Intercept	1.66533E-16	0.25819889	6.44981E-16	1	−0.542455738	0.542455738
Dose	0.04	0.018856181	2.121320344	0.048037528	0.000384634	0.079615366

The parameters of the regression line are in the last table in the column labeled "Coefficients." The intercept is in the row labeled "Intercept" and the slope is in the row labeled "Dose." From those estimates, we can write the regression equation.

$$\hat{p} = 1.7 \times 10^{-16} + 0.04 \cdot \text{Dose}$$

We can use that regression equation to estimate the probability of being cured corresponding to a specified dose. For example, the probability of being cured if a person received a dose of 12 mg is:

$$\hat{p} = 1.7 \times 10^{-12} + 0.04 \cdot 12 = 0.48$$

So, we estimate that there is a 48% chance of being cured if a person receives a dose of 12 mg.

We can use the estimates of the slope and intercept from the output from using the "Regression" analysis tool, but we cannot use the *P*-values in that output to test null hypotheses about the regression equation. Instead, we need to calculate a *z*-value for the test of the null hypothesis that knowing dose does not help estimate the probability of being cured (the omnibus null hypothesis). We can calculate that z-value by using information from the output. We use the regression and total sums of squares and the total degrees of freedom. Those values are in the middle table of the output labeled "ANOVA." The sums of squares are in the column labeled "SS" and the degrees of freedom is in the column labeled "df." From those values, the z-value is calculated as shown in Eq. 3.7.

$$z = \sqrt{\frac{\text{Regression SS}}{\frac{\text{Total SS}}{\text{Total df}}}} \tag{3.7}$$

where

Regression SS   regression sum of squares
Total SS   total sum of squares
Total df   total degrees of freedom.

The next example shows the test of the omnibus null hypothesis for the regression equation in Example 3.22.

Example 3.23 Let us test the omnibus null hypothesis for the regression equation in Example 3.22 allowing a 5% chance of making a type I error.

We begin by calculating a z-value using Eq. 3.7.

$$z = \sqrt{\frac{1}{\frac{5}{19}}} = 1.949$$

To interpret that z-value, we use the Excel function in Eq. 3.3. That P-value is equal to 0.051. Since that P-value is greater than 0.05, we fail to reject the omnibus null hypothesis which states that knowing dose does not help estimate the probability of being cured.

Interim Analysis

4

Abstract

Many types of research take relatively long periods of time to collect the sample. When involved in this type of research, it is often desirable to analyze the data as they accumulate. There is a danger in doing this. This chapter describes that danger and provides methods that can be used for these accumulating data that avoid the danger. There are two approaches described. The first changes the way P-values are interpreted, making it harder to reject the null hypothesis while the data are gathered. The other involves prediction of what the future observations will show and determination of the probability that the null hypothesis will be rejected at the planned end of the study.

4.1   The Problem

In Chap. 1, we discussed the multiple comparison problem. Recall that this occurs when a study includes several hypothesis tests. Another time we encounter multiple hypothesis tests is when accumulating data are analyzed before the planned sample size is reached. This is called *interim analysis*. This occurs very commonly in clinical trials of therapeutic interventions. Clinical trials typically take a long time (years) to complete, because they must wait for eligible patients to be recruited. This is true even of multicenter studies that recruit patients from several clinical sites.

There are a couple of reasons for wanting to end a clinical trial early. One reason is financial. Clinical trials are expensive.[1] If sufficient information to draw a conclusion has been collected before the planned end of a study, a lot of money can be saved by stopping the study early. Other reasons are ethical. If the efficacy of the intervention can

---

[1] A recent estimate is that the median cost of a clinical trial is $41,177 per patient.

© The Author(s), under exclusive license to Springer Nature Switzerland AG 2022   55
R. Hirsch, *Statistical Hypothesis Testing with Microsoft* ® *Office Excel* ®,
Synthesis Lectures on Mathematics & Statistics.
https://doi.org/10.1007/978-3-031-04202-7_4

be established, it is ethically mandated to stop the trail and provide patients with the more efficacious treatment. Another ethical concern is safety. Many interventions have adverse events. If those adverse events outweigh the benefits of treatment, a clinical trial should be stopped, and patients provided the safer treatment.

So, there is strong motivation to analyze accumulating data in a clinical trial or similar study to potentially end a study before the planned end. A problem with doing this is like the problem encountered when a study contains several hypothesis tests. Each time the data are analyzed, there is a chance of making a type I error by rejecting a true null hypothesis. When this is done several times, the overall (experiment-wise) chance of making a type I error can become large. with as few as five analyses, the overall chance of making at least one type I error is 0.2262. This is substantially larger than the test-wise chance of making a type I error of 0.05. The problem is more serious when analyzing accumulating data, because making a type I error results in stopping the study and not analyzing the data further. This means that if one test results in a type I error, all subsequent tests will result in a type I error as well.

4.2   Sequential Analysis

One solution to this problem is to use a lower value of alpha on each one of the hypothesis tests on accumulating data. This is the frequentist solution to the multiple comparison problem. In interim analysis, these methods are called *sequential analysis*. There are two types of sequential analysis. *Open-ended* methods are used when a study has no planned end but continues until a conclusion, to either reject or fail to reject the null hypothesis, is reached. *Close-ended* methods are used when statistical methods like those in Chap. 5 are used to select a sample size. In close-ended studies, the study continues until a conclusion is reached or the planned end of the study is reached.

The first methods proposed for sequential analysis are known collectively as *classical sequential analyses*. One method, developed by Wald, is an open-ended approach. This method is designed to compare two groups with a dichotomous outcome (e.g., cured or not cured). Persons in one group are paired with persons in the other group. The method examines the results when the outcome of each pair becomes known. At that time, the member of the pair with the better outcome is identified and the difference in the pairs with the better outcomes in the two groups is determined. Then that difference is compared to a graph such as in Fig. 4.1.

Another method was developed by Armitage. This is a close-ended method. It is like the open-ended method by Wald in that it examines pairs of patients in two groups by determining the difference in a better dichotomous outcome. It differs from Wald's method in that the study ends if the number of pairs reaches a planned limit. Figure 4.2 illustrates this method.

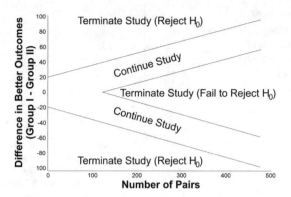

Fig. 4.1 Wald's method for open-ended classical sequential analysis. The difference in better outcomes between two groups is plotted until the value falls in either the terminate and reject the null hypothesis (H_0) areas or the terminate and fail to reject the null hypothesis area. Otherwise, the study continues

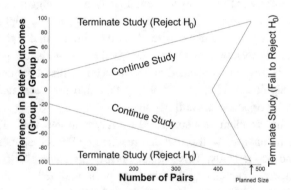

Fig. 4.2 Armitage's method for close-ended classical sequential analysis. The difference in better outcomes between two groups is plotted until the value falls in either the terminate and reject the null hypothesis (H_0) areas or the terminate and fail to reject the null hypothesis area. If we reach the planned end of the study without drawing a conclusion, we fail to reject the null hypothesis

The disadvantages of the classical sequential analysis approaches are that they require the data to be paired and they are designed to examine the data after the results of each pair are known. Pairing patients is difficult to do for it requires assigning similar patients to the groups being compared. This makes patient recruitment more difficult. Analyzing the data after the results of each pair are known is much more frequent than is normally considered necessary. This choice reduces statistical power more than necessary. If the number of analyses were reduced, it would be easier to reject the null hypothesis.

Fig. 4.3 A comparison of three methods for group sequential analysis for four interim analyses and one final analysis. P = Pocock, HP = Haybittle-Peto, OB = O'Brien-Fleming

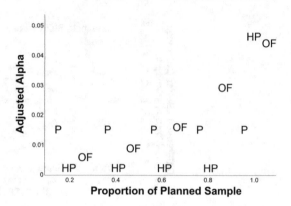

To overcome these disadvantages, several statisticians have developed what are called *group sequential analyses.* In these methods, the alpha is reduced according to the number of times the data are analyzed. All group sequential analyses are close-ended methods.

The most straightforward of group sequential analysis was suggested by Pocock. This approach is like the Bonferroni correction in multiple comparison. Alpha is reduced to a value equal to the desired experiment-wise alpha by dividing that alpha by the number of comparisons. A disadvantage of this method is that the statistical power at the final analysis if the study goes to its planned end is reduced so that a result that would have been statistically significant if the accumulating data had not been analyzed might not meet the criterion for rejecting the null hypothesis.

A solution to this problem was suggested by Haybittle and Peto. They reduced the alpha at each interim analysis well below that required by Pocock allowing a larger alpha at the planned end of the study. This provides nearly the same statistical power at the planned end as if the accumulating data had not been analyzed. Another solution was suggested by O'Brien and Fleming. They set the alpha at the first analysis very low then gradually increased it with each interim analysis. The O'Brien-Fleming approach has nearly the same statistical power at the planned end of the study as the Haybittle-Peto approach. The alphas for each of these methods are compared in Fig. 4.3.

4.3   Stochastic Curtailment

All the methods in sequential analysis reduce statistical power to some degree. An alternative approach that does not reduce statistical power is stochastic curtailment. To see what we mean by stochastic curtailment, let us begin by looking at an example.

Example 4.1 Suppose we are interested in a new topical treatment for arthritis pain. To study this new treatment, we plan to recruit 200 persons with arthritis and give them the new treatment and the leading currently available topical treatment in unlabeled bottles. They are asked to use one of the treatments for a week and then the other treatment for a week and identify which treatment gave them the better relief from pain. After 150 persons had been recruited, 105 preferred the new treatment and 45 preferred the currently available treatment. It was decided that the study could stop recruiting persons and conclude that the new treatment is better. The argument for this was even if the remaining 50 persons all preferred the currently available treatment, more persons among the planned 200 persons would prefer the new treatment.

Example 4.1 shows what is called *deterministic curtailment*. With deterministic curtailment, it is certain that the outcome will go a particular way. What if after the results from 150 persons were known, 90 persons prefer the new treatment. It could be argued that the study could stop and conclude the new treatment is better since it is very unlikely that 40 or more of the remaining 50 persons would prefer the currently available treatment since so far 40% of the persons preferred the currently available treatment. This is an example of *stochastic curtailment*. With stochastic curtailment the result is not certain, but it is very likely.

Stochastic curtailment involves predicting what would happen at the planned end of the study and considering the chance that the result would be statistically significant. That prediction is usually made considering one of three assumptions. One assumption is that the current trend will continue. In the example, the current trend is that 60% of the persons prefer the new treatment. If 60% of the remaining 50 persons prefer the new treatment, that implies 30 (on the average) of the remaining persons will prefer the new treatment. We will consider the other possible assumptions later on, but for now let us assume that the current trend will continue.

It is not enough to consider what the estimate is likely to be. We also must consider what the appropriate hypothesis test will show. For a paired study like the one in the example, the hypothesis test involves calculation of a z-value. That z-value is calculated as shown in Eq. 4.1.[2]

$$z = \frac{p - \theta}{\sqrt{\frac{\theta \cdot (1-\theta)}{n}}} \tag{4.1}$$

where

p observed proportion
θ proportion in the null hypothesis (0.5)
n sample's size.

The next example shows the expected results if the current trend were to continue.

---

[2] This is the same as Eq. 3.2 in Chap. 3. It is the method for the paired z-test.

Example 4.2 For the study of a new pain relief medication compared to the leading available medication in Example 4.1, let us assume that after 150 of 200 persons are observed, 90 prefer the new treatment. What results would we expect to observe if the current trend were to continue?

So far, 90 of 150 persons prefer the new treatment. That is 60% (90/150 = 0.60) of the persons observed so far. If that trend were to continue, 30 of the remaining 50 persons would also prefer the new treatment. That means that, if the study were to go to its planned end and recruit 200 persons, we would expect 120 (90 + 30) persons would prefer the new treatment if the current trend were to continue. That means 60% (120/200 = 0.60) of the persons at the planned end would prefer the new treatment.

The null hypothesis in a paired preference study such as this is that half of the persons will prefer one thing and half will prefer the other. In other words, there is no preference. To test that null hypothesis, we use Eq. 4.1. In the current example, we would expect to get the following z-value after 200 persons are observed if the current trend were to continue.

$$z = \frac{0.6 - 0.5}{\sqrt{\frac{0.5 \cdot (1 - 0.5)}{200}}} = 2.828$$

To find the two-tailed P-value corresponding to that z-value, we use the Excel function in Eq. 3.3. That P-value is 0.0047. Since that P-value is less than 0.05, we can reject the null hypothesis if the current trend were to continue, and we were to observe these results.

So, we expect a P-value of 0.0047 at the planned end of the study if the current trend were to continue. But this is only what we would expect to get on the average. Chance will influence the remaining observations. Even with the current trend, just by chance we might get 29 or 31 persons preferring the new treatment or even a more extreme result. To take this role of chance into account, we calculate the *conditional power*. Conditional power is the probability we will be able to reject the null hypothesis at the planned end of a study under the condition that some assumption is true. The assumption we are considering is that the current trend will continue. To get the conditional power, we need to calculate a z-value. Equation 4.2 shows that calculation.

$$z_{CP} = \frac{z_{alpha} - z_{assumption}}{\sqrt{\frac{n - n_t}{n}}} \tag{4.2}$$

where

z_{alpha}	z-value corresponding to alpha (equal to 1.96 for a two tailed alpha of 0.05)
$z_{assumption}$	z-value expected at the planned end if some assumption is true
n	number of observations at the planned end
n_t	number of observations at the time of the analysis.

The conditional power is the complement of the probability (1-probability) of getting a z-value as small or smaller than the z-value calculated in Eq. 4.2. We determine that probability in Excel with the function in Eq. 4.3.

$$= 1 - \text{NORM.S.DIST}(z_{CP},\text{TRUE}) \tag{4.3}$$

The next example shows calculation of conditional power.

Example 4.3 For the expected results in Example 4.2, let us calculate the conditional power. We begin by calculating the z-value in Eq. 4.2

$$z_{CP} = \frac{1.96 - 2.83}{\sqrt{\frac{200-150}{200}}} = -1.737$$

Then, we use the Excel function in Eq. 4.3 in Excel to determine the conditional power. That conditional power is 0.9588. That means that there is a 95.88% chance of rejecting the null hypothesis at the planned end of the study if the current trend were to influence the remaining observations. If that is considered large enough, the study could be curtailed at this point. If it is not large enough, the study could be continued and reanalyzed after a few more observations have been made.

Assuming the current trend will continue is the most likely assumption of what will influence the remaining observations, but often we want to be more conservative. Another assumption we could make is that the null hypothesis is true and will influence the remaining observations. The next example considers that assumption.

Example 4.4 If the null hypothesis influences the remaining observations, that means that half of the remaining 50 persons will prefer the new treatment. So, at the planned end of the study, we expect 115 (90 + 25) persons to prefer the new treatment, on the average. Then, the proportion preferring the new treatment would be 0.575. The z-value we would expect for that proportion is calculated using Eq. 4.1.

$$z = \frac{0.575 - 0.5}{\sqrt{\frac{0.5 \cdot 0.5}{200}}} = 2.121$$

The P-value associated with that z-value is 0.0339. If the null hypothesis influenced the remaining observations and we were to observe these results, we would be able to reject the null hypothesis. This is the result we would get on the average. To take chance into account, we need to calculate conditional power. To begin, we use Eq. 4.2.

$$z_{CP} = \frac{1.96 - 2.121}{\sqrt{\frac{200-150}{200}}} = -0.323$$

From that result, we use the Excel function in Eq. 4.3 to determine the conditional power. That conditional power is 0.627. That is probably not large enough to stop the study at this point and reject the null hypothesis.

The biggest disadvantage of stochastic curtailment is that there are no rules of thumb for how large a conditional power should be to curtail a study and reject the null hypothesis. So, it remains a judgement call depending on the consequences of being wrong. This is a subjective judgement.

There are two important advantages of stochastic curtailment. One of these is the accumulating data can be analyzed as often as a researcher wants without affecting statistical power. This is not true of sequential analysis.

Another advantage is that stochastic curtailment can be used to stop a study because it is unlikely to result in a statistically significant result. This means a researcher need not continue a futile study. When this is the goal of stochastic curtailment, a third assumption is most often used. That is the assumption that the remaining observations will reflect the threshold of importance.[3] Stopping a study for futility is illustrated in the next example.

Example 4.5 Suppose in the study of a new analgesic medication for arthritis described in Example 4.1 we were to observe 80 of 150 persons preferring the new medication. At this point, we might wonder if it is worthwhile to continue the study. To decide, we can use stochastic curtailment.

Suppose we believe the new treatment will be worthwhile if at least 55% of the persons with arthritis prefer it. Making the assumption that this threshold of importance will influence the remaining observations, we would expect 27.5 of the 50 persons not yet recruited to prefer the new medication on the average. Then, at the planned end of the study we would have 53.75% of the persons preferring the new treatment. At that point, testing the null hypothesis that half of the persons prefer the new treatment in the population yields the following result (from Eq. 4.1):

$$z = \frac{0.5375 - 0.5}{\sqrt{\frac{0.5 \cdot 0.5}{200}}} = 1.061$$

This z-value is associated with a P-value of 0.289. With that P-value, we cannot reject the null hypothesis, but it is possible by chance that sufficient persons of the remaining 50 will prefer the new treatment to reject the null hypothesis. To consider this, we calculate the conditional power. To do that, we begin by using Eq. 4.2.

$$z_{CP} = \frac{1.96 - 1.061}{\sqrt{\frac{200 - 150}{200}}} = 1.799$$

---

[3] This is the same as the minimal detectable difference used in sample size planning discussed in Chap. 5.

The conditional power is found in Excel using the function in Eq. 4.3. That conditional power is equal to 0.036. Now the researchers must decide if this is a small enough chance of a statistically significant result to stop the study at this point or whether they are willing to continue the study with only a 3.6% chance of success.

Planning the Sample's Size

5

Abstract

In Chap. 2, we examined various methods that can be used to select the individuals that will be in a sample. Another aspect of taking a sample is to decide how many observational units the sample should contain. What we would like to do is to take a sample large enough to allow rejection of false null hypotheses. In other words, our sample should have sufficient statistical power. On the other hand, collecting data for large samples is costly. Therefore, we do not want to take a larger sample than is necessary to meet our needs.

5.1 Studies with No Independent Variable

As we learned earlier, independent variables are predictor variables. We will first consider datasets in which there is no predictor variable.

Although there are as many methods of estimating the required sample's size as there are methods to analyze data, the basic principles of those methods are the same. First, we will examine those principles for datasets with no independent variable and a dichotomous outcome (dependent) variable. Then, we will see how those principles apply to other sorts of analyses.

The first step in estimating the required size of a sample is to determine what statistical method will be used to analyze the data obtained from that sample. This is illustrated in the next example.

Example 5.1 Suppose we are interested in studying a new drug designed to treat pain associated with arthritis. In this study, we will select sequential patients seen at a particular clinic (a convenience sample) who have advanced osteoarthritis. These patients will be

© The Author(s), under exclusive license to Springer Nature Switzerland AG 2022
R. Hirsch, *Statistical Hypothesis Testing with Microsoft* ® *Office Excel* ®,
Synthesis Lectures on Mathematics & Statistics.
https://doi.org/10.1007/978-3-031-04202-7_5

randomized to receive either the experimental drug or standard treatment for two weeks. Then, each patient will receive the other treatment for another two weeks. At the end of the four-week period, patients will be asked which treatment gave greater relief from pain. First, we need to consider what method of analysis we will use to test the null hypotheses that there is no preference for one of the two treatments over the other.

This study makes evaluation of both treatments on each patient. Thus, this is a paired study and the data of interest to us will be the proportion preferring the new treatment. Using the flowcharts in Chap. 3, we find that we can test the null hypotheses by using a paired z-test.

In Chap. 3, we learned that the paired z-value is calculated as follows[1]:

$$z = \frac{p - \theta}{\sqrt{\frac{\theta \cdot (1-\theta)}{n}}} \tag{5.1}$$

where

p observed proportion
θ proportion in the null hypothesis (0.5)
n sample's size.

If we examine the equation used to test hypotheses, we can see that it contains n, the sample's size. Thus, to estimate the sample's size we will need for our planned study, we can begin by algebraically rearranging that equation to solve for n (Eq. 5.2).

$$n = \left(\frac{z \cdot \sqrt{\theta \cdot (1 - \theta)}}{p - \theta} \right)^2 \tag{5.2}$$

Examination of Eq. 5.2 reveals that, to estimate the sample's size we will need to provide a z-value, the sample's proportion preferring the new treatment (p), and the hypothesized population's proportion preferring the new treatment (θ). We will consider how we can provide values for each of these one at a time.

First, let us think about what the z-value is doing in Eq. 5.2. In hypothesis testing, we would use Excel to find a corresponding P-value that we would compare to alpha, the chance of making a type I error. In estimating the required sample's size, that z-value should be selected to reflect the alpha we will use in statistical hypothesis testing (we will usually select z to be equal to 1.96 to reflect a two-tailed alpha of 0.05). Since this z-value usually represents a two-tailed alpha, we often give it the subscript of alpha divided by 2 ($z_{\alpha/2}$).

---

[1] This is Eq. 3.2.

We have two more values that we have to consider in Eq. 5.2. One of these is the population's proportion preferring the new treatment. This value is specified in the null hypothesis to be equal to 0.5.[2] The other is the sample's estimate of the proportion preferring the new treatment.

We will worry about providing a value for the sample's estimate of the proportion in a moment but let us first think about what Eq. 5.2 does. It allows us to estimate the sample's size that would be required to reject the null hypothesis for a given value of alpha. Recall from Chap. 1 that alpha is the probability of rejecting the null hypothesis given that the null hypothesis is true. Thus, Eq. 5.2 estimates the sample's size that would be required only under the condition that the null hypothesis is true.

What if the null hypothesis is false and, instead, the alternative hypothesis is true? We can estimate the required size of the sample under the assumption that the alternative hypothesis is true using an equation similar to Eq. 5.2 but based on the alternative, rather than the null, distribution (Eq. 5.3)

$$n = \left(\frac{z_\beta \cdot \sqrt{\theta_A \cdot (1 - \theta_A)}}{p - \theta_A} \right)^2 \tag{5.3}$$

where

z_β a z-value representing the chance of making a type II error (beta)
θ_A proportion in the alternative hypothesis
p observed proportion.

Notice that the z-value representing the probability of making a type I error in Eq. 5.2 is two-tailed, but the z-value representing the probability of making a type II error in Eq. 5.3 is one-tailed. The reason for this distinction is that, when we consider the null distribution (Eq. 5.2) we have no way of knowing on which side of the proportion of that distribution (in this case, the proportion is equal to 0.5) the alternative distribution lies (i.e., whether the proportion will be less than or greater than 0.5). Equation 5.3, however, concerns the alternative distribution. When we consider the alternative distribution in Eq. 5.3, we assume that we know on which side of the null distribution the alternative distribution lies. In fact, we need to specify a particular value for the proportion of the alternative distribution in Eq. 5.3.

Now, we have two equations that give us the required number of observational units in the sample. One equation addresses the type I error (Eq. 5.2) and one addresses the type II error (Eq. 5.3). What we need to do next is to find the size of the sample that satisfies

---

[2] The null hypothesis is that there is no preference between the two things considered. No preference means that the probability of preferring either thing is 0.5.

both equations.[3] The resulting equation is:

$$n = \left(\frac{\left[z_{\alpha/2} \cdot \sqrt{\theta \cdot (1 - \theta)} \right] - \left[z_\beta \cdot \sqrt{\theta_A \cdot (1 - \theta_A)} \right]}{\theta_A - \theta} \right)^2 \qquad (5.4)$$

The process that led to Eq. 5.4 abolished our need to guess what estimate of the proportion we expect to obtain in our sample (p). It left us with four quantities to specify, however. Two of these ($z_{\alpha/2}$ and z_β) are obtained from Excel to reflect chosen probabilities of making an error (type I or type II) in hypothesis testing. The other two quantities reflect the values of the population's proportion. One of these is the value of the population's proportion according to the null hypothesis (θ). This value is obtained by considering the null hypothesis that will be tested. With the null hypothesis that there is no preference, θ is equal to 0.5.

The other value that is required to use Eq. 5.4 is the value of the proportion in the population if the alternative hypothesis is true (θ_A). If the population's proportion actually is equal to the value of θ_A we use in Eq. 5.4, the probability that the planned sample will result in rejection of the null hypothesis is equal to the complement of the probability of making a type II error represented by z_β in the equation. If the actual population's proportion is further from the value specified in the null hypothesis than the value of θ_A chosen to be used in Eq. 5.4, then the probability that the planned sample will result in rejection of the null hypothesis is greater than the complement of the probability of making a type II error represented by z_β (ie, the statistical power). For this reason, the value chosen for θ_A is referred to as the *minimum detectable value*. If the population's value is at least as far from the null value as is the minimum detectable value, then the probability of obtaining a sample for which the null hypothesis will be rejected is at least as great as the complement of beta (ie, the statistical power).

We will look at an example of how we can make these estimates and use Eq. 5.4 to determine the sample's size that is required for a study in just a moment, but there is one part of using that equation that we should discuss first. This is the fact that the two z-values in Eq. 5.4 will always have opposite signs. To help us understand what is happening, let us look at the null distribution and two possible alternative distributions (Fig. 5.1).

Notice in Fig. 5.1 that when the alternative distribution is to the right of the mean of the null distribution, the cutoff between the null and alternative distributions corresponds to a positive standard normal deviate (z-value) in the null distribution and a negative standard normal deviate in the alternative distribution. When the alternative distribution

---

[3] We can do that and get rid of the necessity of guessing the value we will get for the sample's estimate of the proportion at the same time. First, we algebraically rearrange Eqs. 5.2 and 5.3 so that both have the sample's estimate of the proportion on the left-hand side of the equal's sign. Then, we make those two rearranged equations equal to each other. Finally, we rearrange the combined equations once again to solve for the sample's size.

Fig. 5.1 The null distribution
and two alternative
distributions each on its own
standard normal scale

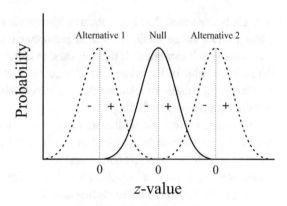

is to the left of the mean of the null distribution, the cutoff corresponds to a negative
standard normal deviate in the null distribution and a positive standard normal deviate in
the alternative hypothesis. Thus, the two standard normal deviates in Eq. 5.4 always have
opposite signs.

Now, we are ready to look at an example of how we can estimate the sample's size
that will be required in a study.

Example 5.2 In Example 5.1, we thought about planning a clinical trial of a new drug
designed to treat arthritis pain. We discovered, in that example, that the method that we will
use to analyze the data from this study will be a preference study testing the null hypothesis
that the proportion in the population preferring the new drug is equal to 0.5.

Now, suppose that we are willing to take a 5% chance of making a type I error (rejecting
the null hypothesis when it is true) and that the alternative hypothesis that will be used in
statistical inference is that the proportion is not equal to 0.5. Also suppose that we are willing
to take a 10% chance of failing to reject the null hypothesis when the alternative hypothesis
is false (i.e., beta is equal to 0.10). Imagine that the drug would be of interest clinically only
if it were associated with at least a preference of 0.6. How many patients should we plan to
recruit for this study?

The first thing that we notice about this problem is the amount of information that is
required to estimate the sample's size for a planned study. First, we need to decide on
the basic study design. This allows us to select the statistical procedure that will be used to
analyze the data we will collect. Then, we need to specify the null and alternative hypotheses
and the probabilities of making both type I and type II errors that we will tolerate during
testing statistical hypotheses. We will use this information to select the z-values that will
be used to estimate the required sample's size. Finally, we need to select the minimum
preference we would like to detect.

To specify the value that will be used as the proportion of the alternative distribution,
we have suggested the smallest value important to clinical use of the new drug. This is a
very sensible way to choose a value for the alternative distribution. If we choose a sample's

size that is large enough to reject the null hypothesis with a given probability when the true value is equal to the smallest value of importance, then we can be even more certain of rejecting the null hypothesis if the true value is larger than this. If the true value is less than the smallest value of importance, we are not terribly concerned with the fact that we will have a low probability of rejecting the null hypothesis. That is because we are not likely to be missing an important finding.

We learned in Example 5.1 that we use a paired z-test to analyze these data. Thus, we will use Eq. 5.4 to estimate the required sample's size. Before we use Eq. 5.4, we need to use Excel to find the z-values that correspond to an alpha of 0.05 and a beta of 0.10. To do that, we use the Excel function "=NORM.S.INV(p)." The "p" in that function is the probability in the upper tail of the standard normal distribution.

The z-value that represents the type I error should correspond to a probability equal to $\alpha/2$ in each tail since the alternative hypothesis is two-tailed. For a probability of making a type I error equal to 0.05 (i.e., 0.025 in each tail), the corresponding z-value is equal to 1.96.[4] As we discussed previously, the z-value representing the type II error probability should correspond to an area equal to beta in one tail. Since the alternative distribution is greater than (i.e., to the right of) the values from the null distribution, this z-value will be negative. Thus, we should use a value of -1.28 to represent the type II error.[5]

Now, we are ready to estimate the required sample's size by substituting these values in Eq. 5.4:

$$n = \left(\frac{\left[1.96 \cdot \sqrt{0.5 \cdot 0.5}\right] - \left[-1.28 \cdot \sqrt{0.6 \cdot 0.4}\right]}{0.6 - 0.5} \right)^2 = 258.3$$

Thus, we should plan to recruit 259 patients for this study.[6]

If we are planning a sample with no independent variable and a continuous dependent variable, there is an additional value that requires our attention: the standard deviation[7] in the population[8] (Eq. 5.5).[9]

$$n = \left(\frac{\left(z_{\alpha/2} - z_\beta\right) \cdot \sigma}{\mu_A - \mu} \right)^2 \tag{5.5}$$

---

[4] The function that yields this result is "=NORM.S.INV(0.025)."

[5] This result comes from the function "=NORM.S.INV(1-0.10)."

[6] The usual practice in estimating the size of a planned sample is to round the result of the calculation up to the next integer.

[7] The standard deviation tells us how spread out the data values are.

[8] We are using the population's standard deviation, so we can use the standard normal distribution, rather than Student's t distribution.

[9] Equation 5.5 was derived using the same logical process that led to Eq. 5.4.

where

$z_{\alpha/2}$ z-value corresponding to the chance of making a type I error
z_{β} z-value corresponding to the chance of making a type II error
σ standard deviation in the population
μ_A the mean in the population according to the alternative hypothesis
μ the mean in the population according to the null hypothesis (usually equal to zero).

As we have seen, the process of estimating the required sample's size involves specifying values for a number of different components of the calculation. This is often the most troublesome part of sample's size estimation and leads to uncertainty about the calculated sample's size because of a lack of confidence in the accuracy of some of the specified values. The solution to this uncertainty is to calculate estimates of the required size of a planned sample using several different values for the component about which we are most uncertain. These calculations are often organized graphically in what is known as a *power curve*. A power curve usually plots the statistical power[10] of a planned statistical test on the ordinate (Y-axis) and the required size of the sample on the abscissa (X-axis). Several lines are plotted on the power curve: each one corresponding to a different value of the component of the sample's size calculation about which we are uncertain. The next example shows how a power curve might be used in planning a study.

Example 5.3 Suppose we are interested studying a new drug designed to lower serum levels of low-density lipoprotein (LDL). In this study, we will select sequential patients seen at a particular clinic (a convenience sample) who have elevated levels of LDL. These patients will be randomly assigned each to receive either the new drug or the standard treatment for a 30-day period and, at the end of those 30 days, they will be assigned to receive the other treatment for another 30 days. At the end of each period, LDL will be determined. Because both treatments are assigned to the same persons, this is a paired study and has no independent variable. Imagine that the new treatment would be clinically important if it lowers LDL by at least 5 mg/dL.

Suppose our best guess for the standard deviation of the difference between two LDL measurements comes from a pilot study and is 8 mg/dL. Even so, we want to consider other guesses (4, 6, 10, and 12 mg/dL) to see how much variation in this guess affects the sample's size. To construct the power curve, we calculate the sample's sizes corresponding to those standard deviations and the z-values representing various probabilities of making a type II error using Eq. 5.5. The following power curve displays the results of those calculations.

---

[10] Recall from Chap. 1 that statistical power is the complement of beta, the probability of making a type II error.

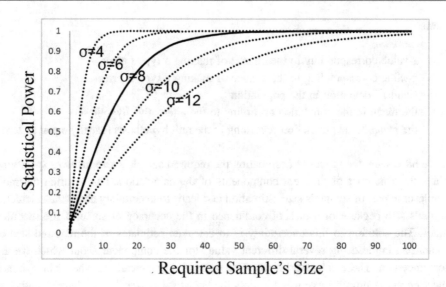

Required Sample's Size

Examination of the power curve shows us that a sample's size of 30 gives us about 90% power with our guess of 8 mg/dL for the standard deviation. That is a reasonable value for the power of the planned study. Further examination, however, reveals that the power will be about 50% if the standard deviation is as large as 12 mg/dL for a sample of 30 patients. On the other hand, we can see that the power will be about 90% if we plan to take a sample of about 65 patients even if the standard deviation is as large as 12 mg/dL. Thus, we should plan a sample of about 65 patients if we believe that the standard deviation of the data in the population could be as large as 12 mg/dL and we want to have a probability of avoiding a type II error equal to 0.90.

5.2   Studies with One Independent Variable

When we have an independent variable, the method of analysis we use is determined, in part, by the type of data represented by the independent variable (see the flowcharts at the beginning of Chap. 3). Since the methods used to estimate the size of the sample that is required are derived from the methods used to analyze the data, we can expect that those methods are different for different types of data represented by the independent variable. Let us begin our examination of size estimation by considering a nominal independent variable.

When we have an independent variable that represents nominal data, the nominal independent variable separates the dependent variable values into two groups. Those two groups might contain the same number of observational units, or they might have different numbers of observational units. Thus, when we have a nominal independent variable, we have another issue that we need to consider when estimating the required size of the sample. That issue is how the total number of observational units in the sample will be allocated between the groups. To specify that allocation, we calculate the number of observations we will have in one group and represent the number of observations in the other group using the *sampling ratio*. This sampling ratio is equal to the number of observational units in group 2 divided by the number of observational units in group 1.

$$R = \frac{n_2}{n_1} \tag{5.6}$$

Equations 5.7 and 5.8 show how this sampling ratio is used in calculation of the required sample's size for a nominal (Eq. 5.7) and continuous dependent variable (Eq. 5.8).[11]

$$n_1 = \left(\frac{\left(z_{\alpha/2} \cdot \sqrt{\theta \cdot (1 - \theta) + \frac{\theta \cdot (1 - \theta)}{R}} \right) - \left(z_\beta \cdot \sqrt{\theta_{A_1} \cdot (1 - \theta_{A_1}) + \frac{\theta_{A_2} \cdot (1 - \theta_{A_2})}{R}} \right)}{(\theta_{A_1} - \theta_{A_2})} \right)^2 \tag{5.7}$$

$$n_1 = \left(\frac{(z_{\alpha/2} - z_\beta) \cdot \sigma}{\mu_{A_1} - \mu_{A_2}} \right)^2 \cdot \left(1 + \frac{1}{R} \right) \tag{5.8}$$

For a simple random sample, the sampling ratio should reflect the frequencies of the two groups in the population. For a stratified sample, the sampling ratio is chosen by the researcher. The most efficient (from a statistical point of view) division of observational units among strata is when each stratum contains the same number of observational units. A different sampling ratio might be used, however, if it is much more difficult to obtain observational units in one group compared to another group.

Now, let us look at an example of how the sampling ratio affects the required size of a sample with a continuous dependent variable and a nominal independent variable.

Example 5.4 We are planning a study of jaundice associated with two different types of liver disease. Suppose that type A liver disease is twice as common as is type B liver disease. Further suppose that we suspect that type B disease is associated with a mean indirect plasma bilirubin level that is approximately 0.25 mg/dL higher than the mean indirect plasma bilirubin level in type A disease. From previous research, we expect the standard deviation of indirect plasma bilirubin values in the population to be equal to 0.20 mg/dL.

We plan to test the null hypothesis that the difference between mean bilirubin levels for persons with these two types of liver disease is equal to zero in the population versus the alternative hypothesis that the difference is not equal to zero. We will randomly select the patients for this study from medical records in 20 clinics that have agreed to participate in

---

[11] Equations 5.7 and 5.8 were derived using the same logical process that led to Eq. 5.4.

the study. Since type A disease is twice as common as type B disease, our sampling ratio (R) will be 0.5 (two cases of type A for every case of type B). We are willing to take a 5% chance of making a type I error and a 5% chance of making a type II error (i.e., a power of 95%). We find from Excel that a z-value of -1.645 is associated with a one-tailed beta of 0.05. Let us estimate how many observational units will be required for each type of disease.

First, we need to determine the method that will be used to analyze these data once the sample has been collected. Since we have a continuous dependent variable and a nominal independent variable, we will use the independent sample Student's t-test for the difference between means.[12] This tells us that we should use Eq. 5.8 to calculate the required number of observational units for patients with type A disease.

$$n_A = \left(\frac{(1.96 + 1.645) \cdot 0.20}{0.25} \right) \cdot \left(1 + \frac{1}{0.5} \right) = 24.95$$

Thus, we should plan to take a sample containing 25 persons with type A disease. To find the required number of persons with type B disease, we rearrange Eq. 5.6.

$$n_B = n_A \cdot R = 24.95 \cdot 0.5 = 12.47$$

Thus, the sample should contain 13 persons with type B disease. The total number of observational units required in the sample is $25 + 13 = 38$ persons.[13]

To demonstrate how the unequal division of observational units between the groups is less efficient statistically[14] than an equal division between the groups, let us calculate the number of observational units that would have been required if we planned the sample in Example 5.4 with the same number of persons with each type of disease (i.e., a sampling ratio of one).

$$n_A = \left(\frac{(1.96 + 1.645) \cdot 0.20}{0.25 - 0} \right) \cdot \left(1 + \frac{1}{1} \right) = 16.63$$

The number of persons with type A disease required in the sample would be 17. A sampling ratio of one indicates that we would also require 17 persons with type B disease. Thus, the total number of observational units required would be $17 + 17 = 34$, fewer than the 38 required if the sampling ratio were 0.5. From a statistical point of view, it would be more efficient to have the same number of persons in the sample with each of the two types of liver disease. A deviation from a sample ratio of one is justified by the difficulty of obtaining observational units in one of the groups.[15]

---

[12] We know that by using the flowchart at the beginning of Chap. 3.

[13] 25 is 12.47 times two rounded up to the next higher integer.

[14] The greater the degree of statistical efficiency, the smaller the size of the sample that is required to achieve a particular level of statistical power.

[15] Another reason for a sampling ratio not equal to one is the desire to take a simple random sample.

When the independent variable in a sample with one independent variable represents continuous rather than nominal data, we do not have a sampling ratio to reflect the number of observational units in one group relative to the other group of dependent variable values. This does not imply, however, that we do not need to be concerned about the distribution of independent variable values in our planned sample. With a continuous independent variable, we use the standard deviation of the independent variable rather than the sampling ratio to indicate the sample's distribution of independent variable values in calculation of the required size of the sample. If we are planning to take a simple random sample, the standard deviation of the data represented by the independent variable that we use to estimate the required size of the sample is our guess at the value of that standard deviation in the population. In that case, the guess that we use is based on the same processes that are used to guess at the value of the standard deviation of the data represented by the dependent variable. Namely, guesses at the population's values of the standard deviations of both the outcome and independent variables can be based on values reported in the literature or on data collected in a pilot study.

It is also possible for us to be planning to take a stratified random sample with a continuous independent variable. For example, suppose we are interested in looking at changes in diastolic blood pressure (dependent variable) associated with various doses of an antihypertensive medication (independent variable). To do that, we plan to assign (or randomize) a particular number of persons to receive each dose of interest. Since we will determine the distribution of independent variable values in the sample, the planned sample would be a stratified random sample of the continuous independent variable, dose.

When a continuous independent variable is sampled as a stratified random sample, the standard deviation of the data represented by the independent variable that is used in estimation of the required size of the sample should reflect the planned distribution of independent variable values. Unlike when we take a simple random sample, the standard deviation of continuous data represented by the independent variable is not something for which we need to make a guess as part of planning the size of a stratified random sample. Rather, we will know the value of this standard deviation once we have decided on how we will assign independent variable values.

In Chap. 3, we learned that there are two analytic approaches we can take when we have a continuous dependent variable and a continuous independent variable. These are correlation analysis and regression analysis. These two analytic approaches have different assumptions and different interpretations. Even so, it does not matter which approach we take when estimating the required size of the sample. This is because, as we also learned in Chap. 3, testing the null hypothesis that the correlation coefficient is equal to zero is the same as testing the omnibus null hypothesis or testing that the slope is equal to zero in regression analysis. Interpretation of the correlation coefficient as a reflection of the strength of the association between the dependent and independent variables in the population is restricted to situations in which we have a simple random sample, but the equality of the tests of inference remains the same regardless of how the data represented by the independent variable were sampled. Thus, we can use a method based on the correlation coefficient to estimate the required size of the sample even if we are planning regression analysis of data from a stratified random sample.

Equation 5.9 shows how we can calculate the number of observational units that would be required in a sample based on the correlation coefficient.[16] The procedure looks more complex than you might have expected. This complexity is due to the fact that correlation coefficients from all possible samples of a given size can be assumed to come from a Gaussian distribution only if the population's correlation coefficient is equal to zero. That assumption is satisfied in the null distribution but not in the alternative distribution. We must transform[17] correlation coefficients that are not assumed to be equal to zero. The following calculation can be used to estimate the required size of a sample regardless of whether we are planning to use correlation analysis or regression analysis.[18]

$$n = \frac{z_{\alpha/2} - z_\beta}{\frac{\ln\left(\frac{1+\rho_A}{1-\rho_A}\right)}{2}}^2 + 3 \tag{5.9}$$

where

ln   natural logarithm

ρ_A the correlation coefficient according to the alternative hypothesis.

If we are using Eq. 5.9 to estimate the required number of observational units for a regression analysis, it might be easier for us to specify the slope of the regression line rather than the correlation coefficient that we would consider the smallest value to be important and, therefore, the smallest we would like to detect (i.e., ρ_A). The most straightforward way to do this is to calculate a correlation coefficient corresponding to the specified slope and then, use Eq. 5.9 to calculate the required sample's size. The relationship between the correlation coefficient and the slope is shown in Eq. 5.10.

$$\rho = \beta \cdot \sqrt{\frac{\sigma_x^2}{\sigma_Y^2}} \tag{5.10}$$

where

ρ the correlation coefficient in the population

β the slope in the population

---

[16] Equation 5.9 was derived using the same logical process that led to Eq. 5.4.

[17] A transformation is an arithmetical manipulation of data.

[18] Equation 5.9 assumes that the null hypothesis is that the correlation coefficient is equal to zero in the population. This is most often true.

σ_X^2 the variance of independent variable values in the population (square of the standard deviation)

σ_Y^2 the variance of dependent variable values in the population (square of the standard deviation).

Now, let us look at an example showing how we can estimate the required size of a sample for a data set containing two continuous variables.

Example 5.5 Suppose we are interested in investigating the relationship between exposure to a particular toxic chemical (ppm) and nerve velocity (m/s) that will allow us to estimate nerve conduction velocities (dependent variable) associated with specific levels of exposure (independent variable). In planning this study, we wish to estimate the size of the sample we will require to reject the null hypothesis that the slope of the regression line is equal to zero. The alternative hypothesis that we will use in statistical inference is that the slope is not equal to zero, but we would like to detect a slope as small as one. We are willing to take a 5% chance of making a type I error and a 10% chance of making a type II error (i.e., a power of 90%). We plan to take a simple random sample of persons in the population. In previous studies of a similar population, the standard deviation of exposure levels was found to be equal to two ppm and the standard deviation of nerve conduction velocities was found to be equal to four m/s. For what size sample should we plan?

First, we need to determine what method of analysis we will use to analyze the data in the planned sample. Since we are planning to take a simple random sample that will consist of a continuous dependent variable (nerve conduction velocity) and a continuous independent variable (exposure level), we could use either correlation or regression analysis. Since our interest is in estimating dependent variable values corresponding to specific values of the independent variable however, our interest is in regression analysis.[19] Thus, it is easier for us to specify the slope that we would like to detect rather than the correlation coefficient. Since calculation of the required size of the planned sample is based on a correlation coefficient, we need to calculate the correlation coefficient that would correspond to a slope of one using Eq. 5.10.

$$\rho = 1 \cdot \sqrt{\frac{2^2}{4^2}} = 0.50$$

Now, we can use Eq. 5.9 to calculate the required sample's size.

$$n = \frac{1.96 - (-1.28)^2}{\frac{\ln\left(\frac{1+0.5}{1-0.5}\right)}{2}} + 3 = 37.79$$

---

[19] This is decided using the flowchart at the beginning of Chap. 3.

Therefore, we should plan to recruit 38 subjects for this study. Since we are planning to take a simple random sample, the distribution of subjects among the independent variable values in the sample will be determined by their distribution in the population (and by chance) rather than specified by the study's design.

5.3   Studies with More Than One Independent Variable

The methods for estimation of the required size of a sample with more than one independent variable that are based on the procedures used to analyze these data are quite complex. That complexity is a result of the number of values that must be specified in their calculation. In addition to the values supplied for datasets with one independent variable, we would need to specify values that mathematically describe all the interrelationships among the independent variables if we were to use these methods. This is a difficult task and usually cannot be done with much precision. It is seldom worthwhile to use these complex calculations if the values we specify for the calculations are poor guesses.

The solution to this problem is to use the sample's size estimation techniques described for datasets with one independent variable even though we plan to use a more complex analysis. For example, if we want to estimate the size of a sample in which we will compare four groups in an analysis of variance,[20] we can plan the number of observational units to be included in each of those groups to be the same as the number we would require in each of two groups if the data were to be analyzed with an independent sample Student's t-test. Then, if we include that number of observations in each of the planned four groups, we will have at least as much statistical power in making pairwise comparisons between the groups as we would if we only had two groups to compare.[21]

---

[20] We learned in Chap. 3 that analysis of variance is a method for comparing three or more means.

[21] We will probably have greater statistical power in the more complex analysis because of the greater precision we can expect in estimation of the variance. We expect this greater precision because we have more groups of dependent variable values that can be used to calculate the pooled estimate of the variance.

Appendix

Posterior Testing

When the ANOVA analysis tool is used to analyze a dataset with more than two groups of continuous dependent variable values, the omnibus null hypothesis is tested. The omnibus null hypothesis in ANOVA is that all the means are equal to the same value in the population. Rejection of that omnibus null hypothesis allows acceptance of the alternative hypothesis. The alternative hypothesis is that not all the means are equal to the same value in the population. That only tells us that there is at least one difference. It does not tell us which means are different from each other. To learn that we need to do a posterior test.

There are several posterior tests that have been proposed. Perhaps the most powerful of those is the *Student-Newman-Keuls test*. To perform the Student-Newman-Keuls test, we begin by arranging the means in order of numeric magnitude. Then, we compare the means starting with those furthest apart. If we can reject the null hypothesis that those means are equal to the same value in the population, we can compare the next most extreme means. This continues until we have tested all the pairs of means or we are unable to reject a null hypothesis for a pair of means. If we are unable to reject the null hypothesis that two means are the same in the population, we also fail to reject the null hypothesis for any pairs of means between the two means just compared.

This protocol for comparing pairs of means will become clearer when we look at an example, but first we need to learn how to make those comparisons. To do that, we need to calculate a q-value. The following equation shows that calculation for the null hypothesis that the difference between two means is equal to zero in the population.

R. Hirsch, *Statistical Hypothesis Testing with Microsoft* ® *Office Excel* ®,
Synthesis Lectures on Mathematics & Statistics.
https://doi.org/10.1007/978-3-031-04202-7

$$q = \frac{\overline{Y}_1 - \overline{Y}_2}{\sqrt{\frac{\text{WMS}}{2} \cdot \left(\frac{1}{n_1} + \frac{1}{n_2}\right)}}$$

where

$\overline{Y}_1$ mean in Group 1
$\overline{Y}_2$ mean in Group 2
WMS   within mean square from the ANOVA output
n_1 number of observations in Group 1
n_2 number of observations in Group 2.

Excel cannot help us find a *P*-value for the *q*-value. Instead, we need to use the *critical value* approach to hypothesis testing. In the critical value approach, we compare the absolute value of our calculated *q*-value to a *q*-value that corresponds to alpha (the critical value). If the absolute value of the calculated statistic is equal to or larger than the critical value, we reject the null hypothesis. If the absolute value of the calculated statistic is less than the critical value, we fail to reject the null hypothesis.

Critical values come from a table. A brief table of critical values for *q* follows. These critical values are all for an alpha of 0.05. If you need a more extensive table, consult a statistics text that discusses the Student-Newman-Keuls test.[1] To use this table, find the column that corresponds to the number of means being compared (*k*). The number of means being compared includes the means in the null hypothesis and any means in between the two in the null hypothesis. Then, find the row that corresponds to the degrees of freedom. The degrees of freedom are the within degrees of freedom from the ANOVA output. If the degrees of freedom are not listed, use the next lower degrees of freedom.[2] Where that row and column intersect is the critical value.

Now, let us look at an example of using the Student-Newman-Keuls test to find differences between means following an ANOVA.

---

[1] One such text is: Hirsch RP (2021) *Introduction to Biostatistical Applications in Health Research with Microsoft Office Excel and R.* 2nd Edition. John Wiley & Sons.
[2] Alternatively, you can find the critical value by interpolation.

Table A.1 Critical values of the q-statistic from the Student-Newman-Keuls test for an alpha of 0.05

k	2	3	4	5	6	7	8	9	10
df									
5	3.635	4.602	5.218	5.673	6.003	6.330	6.582	6.802	6.995
10	3.151	3.877	4.327	4.654	5.912	5.124	5.305	5.461	5.599
15	3.014	3.674	4.075	4.367	4.595	4.782	4.940	5.077	5.198
20	2.950	3.578	3.958	4.232	4.445	4.620	4.768	4.896	5.008
30	2.888	3.486	3.845	4.102	4.302	4.464	4.602	4.720	4.824
40	2.858	3.442	3.791	4.039	4.232	4.389	4.521	4.635	4.735
50	2.844	4.423	3.764	4.008	4.196	4.352	4.481	4.593	4.691
60	2.829	3.399	3.737	3.977	4.163	4.314	4.441	4.550	4.646
120	2.800	3.356	3.685	3.917	4.096	4.241	4.363	4.468	4.560

Example A.1 In Example 3.12, we performed an ANOVA for serum cholesterol levels resulting from use of three drugs and a low-fat diet. We got the following output:

Anova: single factor							
Summary							
Groups	Count	Sum	Average	Variance			
Drug A	10	1520	152	565.7777778			
Drug B	10	1600	160	59.11111111			
Drug C	10	1560	156	260.6666667			
Diet	10	1800	180	233.5555556			
Anova							
Source of Variation	SS	df	MS	F	P-value	F crit	
Between Groups	4640	3	1546.666667	5.528196982	0.003157921	2.866265551	
Within Groups	10,072	36	279.7777778				
Total	14,712	39					

From that output, we know that we can reject the omnibus null hypothesis that all four means are equal to the same value in the population because the P-value in the second table is less than 0.05. Now, let us use the Student-Newman-Keuls test to determine which means are different from each other.

We begin by listing the means in order of numeric magnitude. The means are in the column labeled "Average" in the first table labeled "Summary."

Drug A	Drug C	Drug B	Diet
152	156	160	180

Next, we test the null hypothesis that the two most extreme means are equal to the same value in the population. The two most extreme means are for Drug A and Diet. The within mean square is in the second table in the row labeled "Within Groups" and the column labeled "MS." The number of observations in each group are in the first table in the column labeled "Count."

$$q = \frac{152 - 180}{\sqrt{\frac{279.78}{2} \cdot \left(\frac{1}{10} + \frac{1}{10}\right)}} = -5.294$$

We compare that calculated value to the critical value. We can tell there are 36 degrees of freedom because of the entry in the second table in the row labeled "Within Groups' and the column labeled "df." 36 does not appear in Table A.1 so we use the row for the next lower value, 30 degrees of freedom. There are four means involved in the comparison. The critical value is 3.845. Since the absolute value of -5.2936 is larger than the critical value, we reject the null hypothesis that the mean for Drug A is equal to the mean for Diet in the population.

Since Drug A is significantly different from Diet, we can compare the next most extreme means. That indicates we can compare Drug C with Diet and Drug A with Drug B. For Drug C compared to Diet, the calculated q-value is:

$$q = \frac{156 - 180}{\sqrt{\frac{279.78}{2} \cdot \left(\frac{1}{10} + \frac{1}{10}\right)}} = -4.537$$

For Drug A compared to Drug B, the calculated q-value is:

$$q = \frac{152 - 160}{\sqrt{\frac{279.78}{2} \cdot \left(\frac{1}{10} + \frac{1}{10}\right)}} = -1.512$$

The critical value for those comparisons is the one from Table A.1 that corresponds to 30 degrees of freedom and $k = 3$. That value is 3.486. The absolute value of the calculated q-statistic for the comparison of Drug C to Diet (-4.537) is larger than the critical value, so we can reject the null hypothesis that the mean for Drug C is equal to the mean for Diet in the population. The absolute value of the q-statistic for the comparison of Drug A to Drug B (-1.512) is less than the critical value. That implies that we fail to reject the null hypothesis that the mean for Drug A is equal to the mean for Drug B. It also means we fail to reject null hypotheses for all pairs of means between Drug A and Drug B. So, we fail to reject the null hypothesis that the mean for Drug A is equal to the mean for Drug C and the null hypothesis that the mean for Drug B is equal to the mean for Drug C.

Since Drug C is significantly different from diet, we can test the null hypothesis that the mean for Drug B is equal to the mean for diet. The calculated q-value is:

$$q = \frac{156 - 160}{\sqrt{\frac{279.78}{2} \cdot \left(\frac{1}{10} + \frac{1}{10}\right)}} = -3.781$$

The critical value is the value from the table that corresponds to 30 degrees of freedom and $k = 2$. That critical value is 2.888. Since the absolute calculated value (-3.781) is larger than the critical value, we can reject the null hypothesis that the mean for Drug B is equal to the mean for diet in the population.

A convenient way to report the results of Student-Newman-Keuls test is to construct a table in which those means that are not significantly different get a common superscript. For the serum cholesterol data, that table would be:

Drug A	Drug B	Drug C	Diet
152[a]	160[a]	156[a]	180[b]

So, mean serum cholesterol is significantly different between persons who were assigned to follow the low-fat diet and persons assigned to any of the drugs, but the drugs are not significantly different from one another.

Index

Printed in the United States
by Baker & Taylor Publisher Services